DISASTEROLOGY

DISPATCHES FROM
THE FRONTLINES OF THE
CLIMATE CRISIS

Dr. Samantha Montano

PARK
ROW
BOOKS

PARK ™
ROW
BOOKS™

Recycling programs
for this product may
not exist in your area.

ISBN-13: 978-0-7783-8791-6

Disasterology: Dispatches from the Frontlines of the Climate Crisis

First published in 2021. This edition published in 2025 with revised text.

Park Row Books
22 Adelaide St. West, 41st Floor
Toronto, Ontario M5H 4E3, Canada
ParkRowBooks.com

Printed in U.S.A.

For Mom

This is your book too.

Table of Contents

INTRODUCTION

*"At the start of every disaster movie there's
a scientist being ignored."*

—UNKNOWN

IN THE FINAL days of April 1990, my mother went into labor in a town where babies were being born with an increased rate of cancer.[1] It would take years to show the culprit was toxic chemicals from an old coal tar gasification plant that had seeped out of decaying buried drums beneath the town park, where children played under the watchful eye of their pregnant mothers. The women didn't know that while they sat on park benches worrying about scraped knees and twisted ankles, their unborn children were being poisoned.

Like many other communities across the country, our little town of Taylorville in South Central Illinois was on the cusp of a public health crisis. Our mothers had no expectation of the government riding in on white horses to save them, even if they believed they should. My parents, new to town by way of New York City, were viscerally aware of the fight that lay ahead. In Greenwich Village, they had watched the government ignore the AIDS crisis in the '80s—leaving them with no illusion that a small town, suffering a comparably small medical crisis, would garner attention or action. It was into this fight, in the last days of April 1990, that I was born and dressed in a little white onesie with, "TAG: Taylorville Awareness Group," handwritten in Sharpie across my chest.

As she had during the AIDS crisis, my mother joined her neighbors in a battle that they hoped would one day lead to justice. The group, two hundred strong, demanded the Illinois Environmental Protection Agency and the Illinois Department of Public Health test the contaminated park. It was eventually uncovered that CIPS Gas Plant, the owners of the property, had buried fifty thousand gallons of coal tar underground before selling the land. In 1985, unaware of the buried carcinogens, the new property owners began digging on the site and, in the process, released the hazardous materials.

The families who had suffered direct health impacts sued CIPS and were awarded $3.2 million—not much given the legal fees and lifelong healthcare bills. It took years of constant public pressure from TAG, but the site was eventually cleaned up. For their efforts, Taylorville joined the ranks of communities that have banded together to fight for their own health and safety in the absence of government regulation, oversight, and intervention.

As protests continued in Taylorville, my parents moved back East to a town where they would not have to worry about their children living so close to a Superfund site. So despite a dramatic beginning, I grew up in Maine where my weekends were spent in the mountains and my afternoons by the sea. Valuing nature was part of the culture, so when our middle school science teacher told us about climate change—then called global warming—my first thought was the health of the planet, not what it meant for us.

It was a pre-*Inconvenient Truth* world and our teacher's understanding of climate change was rudimentary. She explained that since the Industrial Revolution, people in industrialized countries had released so many greenhouse gasses into the atmosphere that the entire climate of the planet was changing.[2] By 2100 the coral reefs and ice caps could be gone. At twelve years old, something that might happen at the bottom of the ocean and high up in the mountains in over a century was too abstract for me to grasp.

I also did not believe any of it would actually happen. I understood humans were changing the climate, but I thought we would surely do something to stop it. It was only recently, as I listened to youth climate activist Greta Thunberg, that I realized my healthy skepticism of the seriousness of global warming my science teacher casually presented was rooted in the belief that, if this horrific future was a real possibility, "we wouldn't be talking about anything else."[3] At my most imaginative, I could not envision that climate change could mean anything worse than slightly warmer Maine winters and an ocean that rose a bit closer to the edges of our lighthouses.

Our teachers presented us with simple solutions. We could stop global warming and save the polar bears by making environmentally friendly choices each day. They gave us worksheets with checklists of tasks that would lessen our carbon footprints: recycle, ride our bikes, turn off lights, plant a tree. These alleged solutions reinforced my belief that this was an easily solvable problem. I rarely worried about climate change because when I looked around, everyone was doing the tasks we were told to do. We carried reusable water bottles and many of our parents started driving electric cars. Every classroom had a recycling bin and a sign reminding us to turn off the lights at the end of the day. My parents often discussed the benefits of affixing solar panels to the roof of our house and the local news noted the possibility of developing a wind energy industry in the state. They said climate change would happen *if* we failed to act, but we *were* acting. In my little world, everyone I knew was ticking off the boxes on the checklist.

In high school, Al Gore's *An Inconvenient Truth* premiered, and it occurred to me that perhaps my middle school science teacher had not taught us all the nuances of this impending global crisis. I tried to develop a deeper understanding of the scope and complexity of the problem. I learned that the consequences were not contained to the arctic or the Great Barrier Reef. Climate change threatened everything. I learned some of those consequences were not a century, but just years away. Worst of all, I learned the world was not meeting the carbon reduction targets encouraged by the International Panel on Climate Change (IPCC).[4] All the box checking we had been doing was not translating globally.

In August 2005, the levees broke, and soon afterward I moved to New Orleans to help with the recovery. In New Orleans I found I could not do many of the individual environmental actions that had been second nature to me growing up. I stopped recycling because there were no recycling bins on my college campus. I bought bottled water because the city was often under boil-water advisories. The warped roads and sidewalks, damaged by the recent flooding, made riding a bike an extreme sport. It was not that people in New Orleans did not care about the environment, but that the systems and incentives that had made it easy for us to take these actions in Maine did not exist in Louisiana. The environmentalism I grew up with was one of privilege, void of interrogation of environmental injustice or action related to environmental racism. My family had worried about negotiating tax credits for solar panels on our house, not whether we had a roof to put solar panels on.

I learned more about the complexities of the modern environmental movement and questioned what it meant to be an environmentalist. I could see the limitations of individual behavioral changes, but also felt they were the only thing I could control. Moving to New Orleans and finding I couldn't even do the most basic environmental actions left me feeling completely overwhelmed and lost. So I pulled away from environmentalism and shifted my focus to what I perceived to be the unrelated work of disaster recovery.

In New Orleans, the scope and scale of the damage was unprecedented and the needs endless. The White House was absent, the Federal Emergency Management Agency (FEMA) was ineffectual, and city and state governments were overwhelmed. I spent as much time as I could working with different nonprofits and local groups throughout the city, trying to put the pieces back together, but it was an uphill battle that I often felt like we were losing.

When the oil drilling rig, Deepwater Horizon, exploded on April 20, 2010 off the coast of Louisiana, the environmental movement collided with disaster recovery right before my eyes. Eleven men were dead, oil was gushing unabated out of the ocean floor, and its arrival on the shores of the Gulf Coast was imminent. The BP Oil disaster became the largest marine oil spill in US history.[5] Many of us who had been doing post-Katrina work in New Orleans turned our attention to the coast as we began to understand the full scope of the crisis.

When I look at a calendar now, I can see the crisis unfolded slowly, but in my memory it happened fast. One day a photo of twisting metal blanketed the front page of the *Times-Picayune*, and the next scientists were chartering boats and CNN was livestreaming the gushing oil on an underwater camera. The stories in the *Times-Picayune* grew more alarming by the day, but BP claimed—and the media often repeated—that there was less oil than there was washing ashore and that they had found few dead animals along the coast. This, of course, was a lie made possible by the chemicals BP was dumping in the Gulf to sink the oil out of view.

On one afternoon trip to the Louisiana Coast, I climbed up over a dune and under yellow police tape that was being used to keep people away. The beach was being combed by people in hazmat suits that matched the ones we wore when we gutted flooded, molded homes back in the city. The people in hazmat suits carried rakes along the length of the beach and stopped every so often to pick up a dead animal. They'd put it in a bag and throw it into a truck that followed behind them. I felt an unwieldy anger that day. Not only had BP caused this sweeping disaster but they had been, from the moment Deepwater Horizon exploded, deceiving the public. It was standing on that dune that I fully grasped I wasn't just staring out at an oil disaster—I was standing face to face with the climate crisis.

There was a kind of sick poetry to it all. The oil and gas industry had spent decades tearing up the fragile ecosystem of coastal Louisiana while digging up the fuel that would change the climate. As the seas rise and stronger storms, fueled by climate change, come ashore they cause further degradation to the already weakened ecosystem.[6] Now one of these companies had been caught cutting corners in the name of profit and left the remains of the already battered coastline covered in oil. Hollywood could not have written a more insidious plotline. I didn't know who the hero would be, but the villain was obvious.

At the time I did not understand the extent to which the fossil fuel industry was knowingly perpetuating the crisis by buying politicians and running misinformation campaigns.[7] The narrative I had about climate change excluded the role of the oil and gas industries in creating the crisis by emphasizing the role of the individual consumer. The campaigns for bike riding and solar panels did have environmental benefits, but they also helped to obscure the extent of the crisis.[8] We had been made to feel that the crisis was our fault because we were driving around too much. When really, we were trying to fight a trillion-dollar global machine with energy efficient lightbulbs. Of course we were losing.

While I had been wasting time worrying about eco-friendly water bottles and composting, others within the environmental movement were trying to dismantle the country's fossil fuel infrastructure. They knew that we would not meet the IPCC goals with our individual checklists. It would take getting the United States and the rest of the world off of fossil fuels for us to have a fighting chance. The BP Disaster was everything we had been warned about, and some hoped this could finally be the moment where the vast destruction caused by the oil industry would be acknowledged. If Americans couldn't break up with these companies over carbon emissions, maybe seeing the Gulf Coast covered in oil would change their minds. Volunteers from national environmental groups arrived en masse, some with the mission of using the disaster to change the course of America's energy future.

If there was ever a moment that could shake the industry's stronghold in Louisiana and be a turning point for actual climate action, many environmental activists thought this was it. It seemed reasonable to expect the very people who had spent months cleaning oil off their boats, seeing doctors for unexplained health issues, and unexpectedly living off savings accounts, would turn against the industry responsible.

Months later I was disappointed, though not altogether surprised, to hear that door-to-door campaigns to organize opposition to the oil industry had not been successful. What the good intentioned activists could not overcome was that oil was a primary driver of the state's economy. For many, without the oil industry, there would be no jobs—especially then, with tourism and the fishing industry decimated from the disaster. They were—and are—held hostage.

My understanding of the relationship between the environmental movement, the climate crisis, and disasters began to shift. As I became involved with the BP response and recovery, I started to question my perception that disaster work was something separate from environmentalism. I began to understand the connection between the health of our ecosystems, climate change, and the destruction in New Orleans. My understanding of inequality and injustice began to deepen. I learned how the people who are most affected by disasters are often the people who have the fewest resources to protect themselves and rebuild afterward.

The more I learned, the more I realized how much I did not know. My firsthand experience in Louisiana had made it clear that we were not managing disasters effectively, but I didn't yet know enough to understand the root causes of those problems or how to articulate solutions. One of my professors recommended I go to graduate school to study emergency management. I was skeptical at first. My only knowledge of emergency management was the Federal Emergency Management Agency (FEMA), which, after years of living in post-Katrina New Orleans, was not exactly an agency I wanted to be associated with. My professor assured me emergency management was much more than FEMA and that I should give it a chance. So I packed up a U-Haul and drove from New Orleans to Fargo for an emergency management graduate program at North Dakota State University. There I fell into the world of disasterology and found a framework and a language for understanding the disasters I had been lost in.

Right away I loved studying emergency management and wanted to tell people about the work I was doing, but I started to notice as soon as I said the words *emergency management*, people's eyes glazed over. I would go to a coffee shop or a bar in Fargo and someone would ask me what I do. When I told them what I was studying, they would give me a confused look and say something along the lines of, "so you're a firefighter?" I didn't know how to explain what I did.

It's not like I could hit them with "the profession of emergency management is the managerial function charged with creating the framework within which communities reduce vulnerability to hazards and cope with disasters"[9] or my discipline studies "how humans and their institutions interact and cope with hazards and vulnerabilities, and resulting events and consequences"[10] in the Starbucks line. I needed a better way to quickly articulate what I did, so I started saying I was a disasterologist. No confusing definitions needed—people got that I meant I studied disasters.

The theoretical implications of a field of "disasterology" have themselves been a point of global debate among disaster researchers dating back to at least the 1980s. It's an important discussion within the confines of academia, but to the outside world "disasterologist" just sounds intuitive. I started using the term and it caught on quickly, as other disaster researchers also found utility in describing their work this way. I use "disasterology" in an informal way to describe anyone who does disaster research, regardless of their discipline.

More specifically, I do research within the discipline of emergency management.[11] We study everything related to the disaster life cycle—mitigation, preparedness, response, and recovery. Through this research I have learned the history of disasters and read the empirical research on disasters. I learned how we can better manage them or even prevent them altogether. I also learned that climate and disaster work are inextricably linked. Advocating for hazard mitigation and preparedness are just as much part of the climate movement as protesting oil pipelines.

Years ago, when I first began to understand the urgency of the climate crisis, I struggled to figure out how I could help prevent future climate-driven disasters when I was already standing in the middle of one. How could I justify taking the time to worry over Miami's future when that future had already arrived in New Orleans?

What I did not understand at the time was that we could—and had to—do both. If we do not radically change our emergency management policy and approach to managing disasters, the apocalyptic Hollywood disaster scenes that come to mind when we think about climate change could become real life. My hope is by sharing with you the long and often indirect journey I have taken to understand the true extent of the trouble that we're in, you will see the problems clearly too, and find the courage to take action. Because we have a lot to do.

I wanted to be hopeful on the day the Paris Agreement was signed in 2016.[12] The challenges of bringing together the global community to fundamentally change national policies, economies, and cultures to address a problem primarily driven by just a few countries are immense. There is no precedent for the coordinated effort that is needed to lower emissions across the globe, and it was a good sign that nearly every country in the world recognized action was needed. The agreement is a testament to the fact that although the costs of addressing climate change are vast, the costs of not doing so are worse. In the most basic sense, the Paris Agreement seeks to transform the economic and moral argument for mitigating climate change into a plan.

Any excitement I had about this united front was overshadowed, though, by the knowledge that it was a modest agreement, far from what was actually needed.[13] The Paris Agreement fails to create a mechanism to actually hold oil companies and big offender countries like the United States and China accountable.[14] As Brian Deese, a chief negotiator of the agreement, summarized in 2017, we are "more fucked than we think we are."[15] The crown jewel of international climate policy is not enough. Which is why, as of this writing, while I *hope* we act quickly and decisively on climate change, I do not *trust* that we will.

Climate change involves two separate but related problems. The first is cutting emissions to prevent further changes to our climate, which has been the primary focus of the climate movement. To the extent we continue to be unsuccessful in addressing the first problem, we arrive at the second—addressing the *consequences* of climate change. Accepting the need to manage the consequences of climate change is not conceding that we are doomed. Instead, it is a recognition of the communities that are already living through the climate crisis. It is a recognition that regardless of our actions today, things will get worse before they, hopefully, get better. It is these consequences, and what to do about them, that is the purpose of this book.

The consequences of climate change are all-encompassing. Climate change affects every part of our lives: the jobs we have, the places we live, what we eat, our ability to breathe. These consequences cascade and collide in ways we do not yet even fully understand. The parts we do understand, though, are terrifying—especially the parts about extreme weather. The National Climate Assessment, a compilation and analysis of the latest findings from climate scientists across the world, outlines our current reality and our likely future. The climate is changing and we have begun to see the consequences scientists have warned us of for decades.[16] It is encouraging to read narratives of hope that envision a more sustainable future, but when I look around all I see are the places where time has already run out.

Sea level rise for coastal communities is challenge enough, but climate change is much more than that. The climate crisis looks like 2019's Midwest river flooding, urban flooding in Houston, heat waves so severe that death tolls among vulnerable populations like the elderly are already rising, droughts that bring the end to generations-old farmland, and a wildfire season that is extended year-round. The disaster records are breaking so fast that in the time it's taken me to write this book, I've had to entirely rewrite the following section. Hopefully, it's not already out of date by the time you read it.

In 2018, the Camp Fire became California's deadliest wildfire, killing eighty-five people.[17] In 2020, over four million acres burned across the state (an area the size of Rhode Island and Connecticut combined), more than doubling the state record set in 2018.[18] Hurricane Harvey brought more rainfall than any other tropical storm in recorded US history.[19] The 2020 Atlantic hurricane season broke the record for most named storms on record and tied 1886 and 1985 for the most continental US hurricane landfalls in a single season. The season had three dual landfalls and Louisiana was in the cone of uncertainty a shocking eight times.[20] 2017 became the most costly year in US history with disaster damages totaling around $306 billion.[21] As of February 2021 FEMA was spending $340 million a day.[22]

We need to do major damage control. Hurricanes, flooding, storms, wildfires, and droughts are getting worse, and we need to ready ourselves to manage these acute events while also adapting to a new climate. The world around us is changing fast, and we need to change quickly, too. The disasters of recent years have been devastating, but they are also just the earliest symptoms of the climate crisis. The future we face is even more dangerous.

When envisioning a climate-changed world, many think of apocalyptic scenes in disaster movies. Although this world is possible, our imminent future looks more like *Beasts of the Southern Wild* than *Mad Max*. There are a lot of people and communities who will suffer the consequences of climate change before the world as a whole becomes unlivable.

From the day I was born, I was taught that we are responsible for the protection of our own communities. We cannot guarantee government will protect us or that politicians will prioritize our well-being. In communities that have historically trusted government, mismanaged crises challenge their prevailing narrative of government as protector. In communities that have historically been ignored, harmed, or abandoned by government, disasters reinforced this reality.

In this book, I will take you around the country and introduce you to the places that are already living through climate-related disasters. I will show you how the recovery in New Orleans post-Katrina was driven by local residents and volunteers, and how the Corps of Engineers would have long ago abandoned a little neighborhood on the coast of Maine if local residents had not pressured them into finding a flood solution. I will tell you about the survivors and volunteers that showed up to help during the Tax Day Flood in Texas. I will explain how the federal government failed Puerto Ricans during Hurricane Maria in 2017 and then failed us all during the COVID pandemic.

Activism and local community organizing has long been the route to protecting our communities and the future will be no different. As disaster after disaster exposes the inadequacy of government response, survivors act to prevent their own suffering, seek justice, and rebuild their lives. I will explain how the history of our approach to emergency management has failed to prepare us for our current moment, and how we need to take charge now if we are to minimize the suffering.

We cannot only work on preventing the climate from changing; we also have to work to mitigate the consequences of those changes, at the same time. We can save most of the places and people we love, but we have to act quickly and wisely. We must be brave and precise. There is no room for error. We need to act big, not because there is nothing to lose, but because there is everything to lose.

PART ONE

Recovery: The Second Disaster

"Now, we rebuild."

–THE ROCK, *SAN ANDREAS*

GROWING UP, MY mother kept a little television on our kitchen counter. It had an antenna and played images in black-and-white, long after the time people had black-and-white televisions with antennas. In the morning before school, I would watch the local news fade into the *Today Show*. I stood in front of the little screen, leaning against the granite countertop to get close enough to see the hazy images. The routine of starting each day aware of what was going on around the world made me feel like I was in control, even when the news was bad, as it often was following 9/11.

I was eleven years old when the towers collapsed. The disaster coverage began to unfold live on the second hour of the *Today Show* and by the time I got home from school that day it had evolved into 24/7 coverage for us to watch in our living rooms. The black-and-white kitchen TV had not been big enough to contain that day so we had watched it, in color, on the living room TV. I sat in front of the TV for weeks, watching search and rescue turn into recovery, and eventually co-opted into war. Despite the loss of life, I saw a system at work to fix the disaster itself. I saw droves of help flooding the streets of New York, from volunteers to first responders. I saw the mayor and the president there, at Ground Zero. Although the government failed to prevent the attacks, there was an immediate government response once they happened.

Only four years later, Katrina made landfall along the Gulf Coast. I did not even know there was a hurricane until the levees broke. Helicopters started circling New Orleans, and they sent their footage into my safe and dry house. I watched the Katrina coverage on the black-and-white TV, which made it look like a documentary rather than a crisis unfolding in real time. Everything about Katrina felt far away. It was something happening to people I did not know in a city I had never seen. There was no announcement made over the school loudspeaker. My world didn't stop. Katrina stayed contained in the little black-and-white TV.

I moved on, after a while, when it was no longer the leading story on the morning news. Which is why, months later when the campus minister at my Jesuit high school approached me about going on a service trip to New Orleans to rebuild houses over spring break, I was skeptical. Would there be anything left for us to do? He described a city still in crisis, which added to my skepticism. If they needed so much help, surely they would have made that clear on the *Today Show*. I remained unconvinced, but my friends were going and I was promised an afternoon on Bourbon Street—a tempting offer for a sixteen-year-old. It is impossible to overstate how little I knew at the time about either Louisiana or hurricanes. I didn't exactly understand what a levee was, and I certainly knew nothing about recovering from a disaster but I got on the plane anyway.

So it was that my first foray into the world of disasters happened, not out of benevolence or a need to right an injustice, but rather a want of adventure. It also happened to be for the most famous—or infamous—recovery in modern US history—post-Katrina New Orleans.

My first glimpse of the Lower Ninth Ward was from the window of a fifteen-passenger van filled with other volunteers from my high school. We arrived in New Orleans after nightfall and drove across the city to an elementary school that had been hobbled together into a volunteer camp that we would call home for the next week. The school had been gutted and Sheetrock quickly installed. There were curtains for doors and bathroom stalls were haphazardly made out of plywood. There were dozens of other volunteer groups staying at the camp and we were each assigned various chores to keep it running.

As we drove to the camp, I realized there were houses on either side of us, camouflaged by the night. Many months had passed since Katrina made landfall, but in many places, electricity had yet to be restored. There were no streetlights under which kids played, no porch lights left on for teenagers rushing home to make curfew, no blue-light glow from late-night TV, or twinkling lights on backyard patios. There was darkness, pierced intermittently by sputtering generators and brief passing headlights. As ominous as this darkness was, daylight brought something worse.

In the light—and in color—everything was broken. Debris, covered in toxic flood water, was piled high for miles, rotting in the heat. There was no quiet. The streets were filled with the constant sounds of collapsing buildings, hammers pounding, and saws buzzing. It's the way my skin felt, though, that I remember most. The smell in the air was so strong it sunk into our skin, mixing with the particles of dirt and who-knows-what that filled the air, clinging to our sweat and leaving us covered in a layer of grime.

For miles in every direction, houses lay in pieces, stores sat empty, cars abandoned. There was no mail service, no trash or recycling service, no public transportation, only a shadow of a school system, a fledgling healthcare system, essentially no mental healthcare, and a local government that was more than overwhelmed. People were dead, missing, injured, and displaced. The economy had collapsed. Every part of the city had been impacted. Every facet of modern life, gone. *The* Today Show *really held out on us*, I thought.

I have since learned that the national media generally, and especially as compared to local media, often fails to adequately and correctly capture the complexities of disasters. This isn't just frustrating, it is also dangerous. The media has an important role to play in how we respond to disasters, especially in a situation like New Orleans, when government is failing to meet survivors' most basic needs.

To be fair, even if the national media had done more to continue their Katrina coverage, I don't think I would have been capable of grasping the extent of destruction. It was not until I was there, surrounded by the heaps of molding homes, that I could even begin to understand how much help was needed. We worked mostly in the Lower Ninth Ward, a predominantly Black neighborhood on the edge of New Orleans where 100 percent of homes had been flooded.[23] Many were a total loss, pushed completely off their foundations.

Most Lower Nine residents were still scattered around the country, but the ones who had found a way back were starting to wrap their minds around what it would take to rebuild the homes their families had lived in for decades. Throughout the week, these survivors graciously told us the story of Katrina and the flood. It was not the story I had heard on the *Today Show* or other news programs—likely a combination of my own lack of attention and how coverage was framed. I was at the very start of my education in understanding what had happened and how it had happened. The story of how the levees broke and all that came before and after was heartbreaking, overwhelming, and infuriating. What was more, as I looked around I did not see the same help that I had seen in New York City after 9/11. I did not see government officials walking around or people in official-looking uniforms. All I saw were high school and college students in matching T-shirts wielding hammers and nails. Nothing seemed to look the way I assumed the recovery of a major US city should look.

In the movies, disasters have happy endings. At the end of *San Andreas*, survivors of the earthquakes and tsunami are shown arriving from the smoldering city to the temporary shelters on the hills overlooking San Francisco. The dramatic, life-saving tasks have ended, and officials from the National Guard and FEMA have arrived. Dwayne "The Rock" Johnson stands with his family overlooking the crumbling Golden Gate Bridge and delivers the closing line of the movie, "Now, we rebuild."

Notably, The Rock's character did not ask how they would rebuild or who would pay for it. He did not ask who would be able to rebuild and who would be kept from ever returning. Perhaps Hollywood doesn't ask those questions because they don't come with answers we like. If Katrina had been a movie, the end credits would have rolled as the final buses picked up the last of the evacuees and pulled away from the New Orleans Superdome, signaling the crisis over. In real life though, the crisis of the recovery was only just beginning.

Katrina has had an unusual staying power on the public consciousness. It has become a part of the cultural fabric of America, from the HBO series *Treme* to the backdrop of Beyoncé's "Formation" video and credited as a pivotal event in the emergence of the Black Lives Matter movement.[24] Even so, many Americans moved on remarkably quickly when New Orleans was no longer the lead story on the *Today Show*. If I had not, almost by accident, gone there myself, I would have been among those who had moved on. It would have stayed in that black-and-white TV.

On our trip we met a guy who wore a crisp white cowboy

hat and drove a matching Ford truck. He wasn't from New Orleans, but he had been volunteering for months. We considered anyone who had been there longer than we had an expert, so we clung to his every word. He told us he thought it would take ten years for the city to recover. I remember being shocked. At sixteen years old, a decade is more than half a lifetime. In retrospect, my surprise was naive and his answer too optimistic.

The Lady in the Purple Outfit

WHEN YOU REBUILD a house in the Lower Ninth Ward your tools and supplies can get away from you. They spill out into the street and into neighbors' yards as the day drags on, which is how I met the lady in the purple outfit.

She lived in a FEMA trailer that sat on the edge of her yard and next to the house we were rebuilding. Her house was still standing, but it wasn't livable. The lady in the purple outfit is the first Katrina survivor I remember meeting. She wore a bright purple sweatshirt with matching purple sweatpants, giving her the appearance of a beacon among the construction and piles of debris. She insisted she would rebuild. She was determined. The Lower Ninth was home. New Orleans was home. Our group had some unspent money from our trip—just a few hundred dollars—and we gave it to her. I think she said something about buying a washing machine.

Katrina, like all disasters, uprooted communities along its path physically, socially, and economically.[25] In New Orleans, nearly all buildings were either damaged or completely destroyed. Streets cracked under the weight of toxic flood water, and the already aging water, sewage, transportation, and electrical systems were destroyed. Much of the flood infrastructure, including several levees that had failed, was literally left lying in pieces. The debris from the flood could fill the Superdome at least eleven times.[26]

The flood sent families and friends in different directions. Every aspect of New Orleanians' social lives was disrupted: school and work routines, preferred hair and nail salons, childcare facilities, afterschool activities, nightlife, the postal service, Saints games, Mardi Gras preparations—everything. Most New Orleanians had to physically relocate outside New Orleans, at least while the city was drained. The distance and scattered nature of the exodus strained social networks.

The flood affected the natural environment too. In combination, Hurricanes Katrina and Rita—which came through a month later—caused significant wetland loss along the coast and destroyed over one hundred thousand trees in New Orleans alone.[27] Chemicals from the refineries throughout Louisiana washed into the city, seeping into the ground. A sheen of toxic flood water coated everything and contaminated the soil. Katrina became, at the time, the largest oil spill in the US since Exxon Valdez.[28]

The statistics used to describe a disaster run the risk of sterilizing the pain, but they do provide practical value. We have to quickly quantify the damage because it helps us understand the extent and types of resources a community[29] will need to recover. Coming up with these numbers is a complicated process though, which requires the involvement of many different people and organizations. Individuals informally account for their own damage and more formally for insurance claims. Churches survey parishioners to learn the extent of their congregation's damage. Local governments have to assess the damage to public infrastructure and government buildings. Businesses gauge their physical damage and indirect impacts like disruptions to their supply routes, employees, and customer bases. The result is an imperfect, piecemeal picture of what has been lost. The statistics that reporters lead with are often, at best, underestimations.

The damage I saw in New Orleans was visually and viscerally shocking, but what most captured my attention on that first trip was how much *need* there was throughout the city. Everyone needed something and each person needed something slightly different. Few had the resources to meet those needs themselves.

An overwhelming 1.1 million housing units were damaged,[30] accounting for about 70 percent of available housing in New Orleans.[31] The number is staggering, but it does not tell us what the people who lived in those 1.1 million housing units needed. As I talked with New Orleanians, I learned how that number alone did not tell us how many of those homes were damaged to the point of being unlivable. It didn't tell us how long it would take to repair each home or how much it would cost. It did not tell us if the owners could afford those costs. It did not tell us if the homeowners needed temporary housing until they could move back or if they had found someplace to stay. It did not tell us if they could afford their temporary housing. It did not tell us if their pets could stay there. It did not tell us if it was near their job or their children's school. It did not tell us if the homeowners even wanted to move back. Quantifying the damage is not the same as understanding what survivors need, which is where our focus should be.

It was never the mountains of debris that haunted me, it was the fear of an uncertain future on the faces of survivors. The adrenaline of the crisis ends, weeks turn to months and survivors begin to realize those months will turn into years. They accept there is no going back to the way life was before.

Many New Orleanians we met, including the lady in the purple outfit, asked us to tell our friends and family back home about the state of the city. The outside assistance that had been helping to prop the city up was beginning to fade as New Orleans began to feel the mixed effects of the media's absence. Post disasters, the media's departure may, at first, bring a bit of relief. Most of us do not want to be front-page news. It is also a signal, though, that outside help will begin to slow. This was a daunting prospect in New Orleans considering the work of recovery was still just beginning.

I returned home to Maine in a bit of a haze. Everyone around me was carrying on with their regular lives, seemingly unaffected by the pain and turmoil unfolding to our south. They drove to the grocery store when they needed to and had a safe place to sleep at night. There was no debris in the streets, and no one was worried about whether insurance checks would arrive. I found the casualness with which my friends, family, and neighbors moved through life disorienting. I now felt the fragility of my "normal" life and realized that disaster could strike anywhere, at any moment.

I was unsettled by the ease with which I had, to that point, been able to distance myself from communities in crisis. It was in the damp streets of the Lower Ninth Ward, in the shadow of professed 9/11 patriotism, that any illusion of American greatness and solidarity crashed and drowned. In New Orleans, the United States stood exposed.

When I got home to Maine I was met with questions of, "How bad is it really? Are they doing okay down there?" A week earlier I would have asked the same questions, but now I was annoyed. Obviously it was bad! Of course things weren't okay! Eighty percent of the city had been underwater! Common sense would surely dictate that rebuilding an entire city was a more complex task than a few months of work could achieve. Rebuilding would be a monumental task under the best of circumstances, and the government's response was far from the best. I had strong "why don't you know everything about this thing that I just learned about" energy, in a way that only self-righteous sixteen-year-olds can muster.

There was something else that bothered me too. I could hear now a kind of breathless wonderment in the way they asked. It gave away their voyeurism disguised as concern. Regardless of motivation, the question provided me an opportunity to fulfill the request of the people I had met in New Orleans, but it required me to walk a line I still struggle to, between sightseeing and storytelling. A line where tourism and disaster porn can transform injustice into spectacle. A line between amplifying the experience of survivors and speaking for them.

I struggled to articulate the feeling of being in a destroyed city. I wrote journal entries, Facebook picture captions, class essays, and formal reflection papers where I searched for the exact string of words that would, if not actually capture, at least not trivialize what I had seen. I wrote around what I did not yet have the vocabulary to articulate or the historical knowledge to contextualize. Post-Katrina New Orleans was the manifestation of injustice I had, until then, only read about, and with only a novice understanding of those persistent injustices, I could not find the words to explain my anger.

The only disaster writing I had been exposed to measured destruction as a string of statistics—the percentage of homes destroyed, the number of people who had not yet returned, the number of cars flooded, the number dead and missing. These descriptions were shared to elicit maximum shock and keep us reading. I succumbed to this approach, and found that these numbers surrounded by words could turn faces full of sympathy to shock. I was not satisfied though. I was not looking for my friends and family to be shocked. I wanted them to be angry. More elusive still, I wanted the power to motivate their anger into action. I was incensed by what I perceived as indifference toward the depth of need in New Orleans. When I failed to elicit action from my friends and family, I felt I had failed to accomplish the one thing that had been requested of me—to tell the story of New Orleans.

I wanted people from home, a very white, very northern place, to understand that Katrina was not some unforeseen tragedy, but rather the logical outcome of US history. The story of Katrina is the story of racism, classism, incompetence, corruption, and indifference spun around in a hurricane that left lifeless bodies floating down Main Street. There was no act of God, twist of fate, or freak of nature—it was designed. Katrina was the inevitable conclusion of our past, and a glimpse of our future. Katrina subverts the quintessential American story. It is the antithesis of white picket fences but, still, as American as apple pie.

A Heckuva Job!

THE WORD *DISASTER* was too small to describe the crisis in New Orleans.

When the levees broke, journalists descended on the city. News crews filmed scenes that often contradicted what public officials claimed. Politicians spun through a revolving door of interviews where they placed blame and responsibility everywhere, except on themselves. Anderson Cooper memorably scolded Louisiana senator Mary Landrieu on CNN about the number of dead bodies piling up on sidewalks. The anguish in Cooper's questions was met with rehearsed lines and no meaningful plan of action.

Journalists around the country interviewed displaced New Orleanians who had evacuated to other states and were searching for information on their missing family members and damage to their homes. Journalists in New Orleans talked to people who had been waiting for days at the Superdome and Convention Center for buses to evacuate them from the city. They interviewed people walking through the few remaining dry streets and others who passed by in boats. All recounted disturbing scenes from the parts of the city that were still inaccessible. The anguish in their stories was universal.

The world tried to make sense of what they saw on TV as the scope of flooding and ineffectiveness of the federal response became clear. How could it take days to rescue people from their roofs? Why was it taking buses so long to evacuate the Superdome? What the public was correctly picking up on but did not have the language to explain (including sixteen-year-old Samantha) was that Katrina wasn't a disaster. It was a *catastrophe*.

Catastrophe is the term disaster researchers use to describe the worst of the worst.[32] You know them when you see them. (Think: the 2004 Indian Ocean Tsunami, the 2010 Haiti earthquake, and Hurricane Maria in 2017.) They dominate international news, as famous journalists are flown in to report from the streets for days on end. Doctors descend to treat the injured as local healthcare systems collapse. The death tolls tend to be staggering. What is most important to understand though, is that the way you manage a disaster isn't the same as the way you manage a catastrophe. Catastrophes are not just big disasters. They are something different, something more complicated.

As Katrina crossed Florida and quickly gained strength over the warm waters of the Gulf, residents in Mississippi, Alabama, and Louisiana braced for the storm. Navigating conflicting and confusing evacuation orders, some left town, driving north to seek shelter. There were others who refused to leave. They were determined to ride out the storm at home as they had during Betsy, Camille, and all the other hurricanes for generations. Some, though, had no choice. They would stay behind—despite knowing the danger—because preexisting medical conditions made hours in a hot car more dangerous than a wall of water. Many did not have access to transportation or the financial resources to leave. It was the end of the month, which meant, for many, bank accounts were empty.

The devastation along the Alabama, Mississippi, and Louisiana shorelines was quick and complete. Entire towns were obliterated, no match for the storm surge that rose as high as twenty feet.[33] Houses vanished, leaving behind only their cement foundations. As the water pushed its way toward New Orleans, the decades-old, poorly built, and lightly maintained federal flood infrastructure, meant to keep the city from drowning, fell to pieces.[34] As the levee system around the city failed, people moved to second floors and broke through attic ceilings trying to stay above water that reached a depth of twenty feet in some places. In the end, 80 percent of New Orleans was underwater.[35]

The most immediate need in the city was evacuating the hundred thousand people who had remained throughout the storm and now needed to get out.[36] Those who could waded, swam, and floated through toxic flood water downtown to the Superdome and Convention Center or to one of the interstate overpasses that rose above flooded streets. From there they lined up, in inhumane conditions, often for days, for a bus to anywhere else.

New Orleanians did not just sit around waiting for help. They also organized, helped each other, and advocated for themselves. They commandeered school buses to evacuate their friends and family, they searched for food to feed their children, and rescued elderly neighbors off rooftops. They were joined by some local first responders who assisted with search and rescue, but trained responders were immediately overwhelmed by how many needed help.

The efforts of first responders were further hampered because 16 percent of the New Orleans Police Department did not report for duty during Katrina, which is a rare example of first responders abandoning their jobs during a disaster.[37] Although, even if they all had reported for duty they may not have been much help. There were extensive criticisms over how law enforcement, as a resource, was used. NOPD patrolled the city for looters, rather than the many more pressing needs like search and rescue. NOPD leadership also perpetuated horrible rumors about violence at the Superdome and Convention Center that helped to fuel tensions in the city.[38] Worse still at least eight NOPD officers were involved in the shooting and killing of three unarmed Black men—Henry Glover, James Brissette, and Ronald Madison—in the days after the flood.[39] It took years, but the officers have all since been found guilty.

While Mayor Nagin did fire off impassioned pleas for help in the media his actions—from the timing of the evacuation orders prestorm to his approach of requesting aid from state and federal officials after—have been a point of much scrutiny.[40] He opted to ride out the storm at the Hyatt rather than City Hall and when his cell phone died, no one could reach him. It was immediately clear that Mayor Nagin was overcome by the situation and the city did not have the resources necessary to manage what was unfolding in New Orleans.

A defining characteristic of catastrophes is that local leadership is incapacitated. This may be because local officials are directly impacted themselves, or because they do not have the resources to respond, if not both. This was obviously the case in New Orleans. When this happens, leadership must come from outside the impacted area in order to move the response forward. After the mayor, the governor would have been an obvious choice to be the face of the response. Yet, Louisiana governor Kathleen Blanco seemed stuck between city and federal officials. A key failure of the response was that it took days to fill the leadership vacuum.

As far as New Orleanians were concerned, FEMA was nowhere to be found. During the storm it was reported there was only one FEMA employee in New Orleans—Marty Bahamonde. He stayed at the EOC in City Hall before moving to the Superdome and valiantly reported back to FEMA administrator, Michael Brown, the desperate situation in the city. His reports seemed to make little difference as Brown's responses were void of understanding the urgency of the crisis. Further, Bahamonde's reports were not sufficiently relayed to the White House. There was a near complete communication breakdown among federal officials and as Richard Falkenrath, the Homeland Security Advisor, reflected later that "there was a period of days when we weren't sure who was directing the federal response."[41]

Michael Brown, a college friend of Bush's campaign manager, had been appointed to lead FEMA in 2003 just as the agency was being moved under the newly created Department of Homeland Security[42]. Despite having no prior emergency management experience before his arrival at FEMA in 2001, Brown was confirmed by the Republican-controlled Senate.[43] We then got to watch federal officials learn an extremely obvious lesson in real time: the nation's top emergency manager should probably have some emergency management experience.

Meanwhile, President George W. Bush remained on vacation, inexcusably addressing other nonemergency issues for days before finally forming a federal task force to coordinate the federal response and deploying desperately needed federal resources. Five days after Katrina made landfall, thousands of New Orleanians were still on their roofs while President Bush held a press conference in which he praised Michael Brown with the now infamous line, "Brownie, you're doing a heckuva job!"[44] (He was, decidedly, not doing a heckuva job.)

In case there was any uncertainty in how New Orleanians felt about the federal response, a *Times-Picayune* editorial from September 4 called for every official at FEMA to be fired and "Director Michael Brown especially."[45] On local radio Mayor Nagin reinforced the resentment toward the federal government, saying: "Now get off your asses and do something, and let's fix the biggest goddamn crisis in the history of this country."[46]

It wasn't until General Russel Honoré, who was named commander of Joint Task Force Katrina through the Department of Defense, arrived to coordinate military efforts in New Orleans that a more competent federal action began to take shape. General Honoré took a proactive approach from the start, skirting the rules to pre-position troops in Mississippi, prior to the storm's landfall. In fact, he seemed to be one of the few who understood the severity of the situation even as he faced barriers in being able to act. He put into words what many people were thinking, saying, "We've got a case where we need to save life and limb. We can't wait for a [Request for Assistance] or shouldn't be waiting for one. If there's capability, we need to start moving."[47]

Once in New Orleans, General Honoré found the NOPD driving through the streets with guns out, seemingly more an occupying force than anything else. Memorable video footage shows General Honoré walking through the streets shouting, "Put those damn weapons down. I'm not going to tell you again, goddamn it. Get those goddamn weapons down."[48]

Around the country, there was agreement that the response from the federal government was totally inadequate. One opinion poll in the days after Katrina found 67 percent of Americans, including 50 percent of Republicans, thought President Bush could have done more.[49] Even Bush himself thought he was doing a bad job. Six days into the response he admitted, "many of our citizens simply are not getting the help they need...that is unacceptable."[50] FEMA's response was slow and ineffective, as Brown inexplicably failed to understand the severity of the crisis in New Orleans. In the middle of the two weeks after the storm, Homeland Security Secretary Michael Chertoff finally pulled Brown back from New Orleans and he resigned days later.

Offers of foreign aid in its aftermath underscored the spectacular failures of the federal response. Canadian search and rescue teams were some of the first to arrive in Louisiana and Mexican troops crossed into Texas to feed evacuees. Over the following months, donations were sent from nearly every country in the world. Entire organizations were created in other countries to coordinate volunteer missions to New Orleans, efforts that continued for years into the recovery.

These generous offers of aid sometimes came with criticism or at least questions about how a country like the United States, with such a wealth of resources and an advanced emergency management system, had ended up in this situation. It's important to understand that the United States was overwhelmed by Katrina not because we did not have the resources needed, but because the federal government failed to use them effectively. The response was so egregious that international humanitarian organizations that rarely become involved in US affairs came to Louisiana to assess potential human rights violations.

While formal response agencies, and government more broadly, were at once overwhelmed and absent, regular people didn't hesitate to start helping. As the scope of the flooding became known around the country, help converged on the city. For example, in seeing how many New Orleanians were trapped by floodwaters, people from nearby towns towed their boats to Louisiana. These spontaneous volunteers became known as the Cajun Navy and are believed to have rescued over 10,000 people.[51] (One notable exception to the breakdown in the federal response was the Coast Guard, which rescued around 24,000 people.[52])

In a paper published following the storm, Dr. Enrico Quarantelli, a founding father of disaster sociology, argued Hurricane Katrina and the levee failure was an "almost textbook case of a catastrophe."[53] In a catastrophe, impacts, in many forms, are substantial. Katrina was a catastrophe so bad the dead were not accurately counted (foreshadowing Hurricane Maria in Puerto Rico twelve years later). By best count, 1,833 people[54] were killed during Katrina, although the exact number is still disputed.[55] The majority of victims in Louisiana drowned, indicating their deaths were caused by the levee failure, not just the hurricane. It took two or three weeks to find most of the victims, as the damage was so extensive. Nearly half the deaths in Louisiana were people over the age of seventy-five and over half were Black,[56] confirming that disasters do indeed discriminate. That the death toll was not higher is a testament to the courage of the people who did not wait for government direction or permission to act.

In a disaster, assistance can be pulled in from neighboring towns and states. In the case of a catastrophe, and as demonstrated during Katrina, this becomes more complicated. It wasn't only New Orleans that was impacted and needed help. Katrina's impacts rippled out to neighboring states. The effects reached inland as well, not only in terms of the storm itself but also in the emergency exodus from the city. The region as a whole was overwhelmed as cities like Baton Rouge and Houston found themselves struggling to meet the needs of thousands of evacuees.

In catastrophes, survivors and local first responders are unable to address everyone's needs without outside assistance, so basic human needs go unmet for extended periods of time. Survivors may remain isolated for days or weeks with outside agencies unable to make it into the impacted area. This was true in New Orleans where the support the city could usually rely on from neighboring communities was unavailable. Help had to come from farther away (including internationally), which took time.

Once help did arrive in the city, although it was absolutely necessary, it did further complicate the response. With so many organizations, agencies, and individuals involved, coordination was impossible. Many of these groups were not working from the same plan and had different types of training, if they had any training at all. This kind of improvised response is expected in catastrophes, but it can complicate coordination and communication.

This was ultimately what the federal government failed to understand. When 80 percent of an American city is under up to twenty feet of water, you can't hesitate. Catastrophes require a different type of response from federal agencies because, by their very nature, local and state resources are overwhelmed. It can be assumed local government is not functioning and will likely be unable to request help in the traditional way. In a catastrophe, FEMA cannot wait to act until they hear from local governments—the federal government needs to immediately take on a more substantial role. This is what Quarantelli means when he says leadership must come from the outside during catastrophes.

Katrina wasn't like the other disasters FEMA had responded to, and yet, they showed up with their regular disaster plans, not catastrophe plans. It was like bringing a knife to a gun fight. They were completely outmatched.

Apparently, Kanye West's diagnosis that "George Bush doesn't care about Black people" wasn't considered a complete explanation for the failures of the Katrina response. So multiple congressional investigations, dozens of books, hundreds of reports, and thousands of studies have parsed every minute of the Katrina response, all to determine what went wrong and *why*. The answer can largely be summarized as: decades of policy decisions made at all levels of government, including underinvestment in infrastructure, racist housing policies, backroom development deals, and ill-advised post-9/11 changes to emergency management collided with incompetence, bad communication, and poor planning to create catastrophe.[57]

New Orleanians were rightly furious, because while these factors may explain what happened during the response to Katrina, they do not excuse it. It was well-known that the federally built levee system could fail during such a storm. In fact, a scenario prescient to Katrina was the subject of a government-wide exercise in 2004 called Hurricane Pam. The Bush administration cut funding for the exercise before it was completed.[58] As Hurricane Katrina approached, engineers and meteorologists pleaded with government officials to issue mandatory evacuations earlier than they actually did. Leaders at all levels of government knew that a city underwater was a likely outcome, and still evacuation orders were slow to be issued, publicly funded evacuation options were not made widely available, and adequate life-saving supplies were not strategically stockpiled. Then, once the city was flooded, the federal government doubled down on their denial of responsibility while repeatedly failing to provide necessary help.

Eventually Congress commissioned "The Federal Response to Hurricane Katrina: Lessons Learned," which concluded,

> "Katrina was a national failure, an abdication of the most solemn obligation to provide for the common welfare. At every level—individual, corporate, philanthropic, and governmental—we failed to meet the challenge that was Katrina."[59]

When these reports were written, the "challenge that was Katrina" had only just begun. This extensive criticism was just about the *response* to Katrina, not the *recovery*. Their proclamation of failure, though accurate, was still premature, but certainly indicated how the recovery would unfold. While many thought the government should be doing more, there certainly wasn't much trust in New Orleans that they would. So it seemed that the plan was to see if an entire city could be rebuilt by survivors and volunteers well-fed on red beans and rice.

Understanding the response to Katrina as a failure, particularly a federal failure, is critical, because it provides the context for how New Orleanians felt as recovery began. The disdain for the federal government, and FEMA specifically, was visceral. From the moment I arrived in New Orleans this message of government failure was reinforced by everything I saw and every conversation I had. In fact, the entire reason we were there volunteering was predicated on the belief that the government had failed.

As we walked down Bourbon Street on our promised afternoon in the French Quarter I noticed the souvenir shops were filled with T-shirts mocking the federal government's response. I bought one that read FEMA: Fix Everything My Ass.

Do-It-Yourself Recovery

THE SECOND I got back to Maine I started investigating ways to move to New Orleans. I quickly hatched a plan and just a few months later my return to New Orleans became front-page news.

There was so much work to do in the city and I wanted to help. I decided the easiest way for my parents to agree to this was to go to college in the city while I volunteered. I applied early decision and counted down the days until my high school graduation. A few months before I left Maine to move to New Orleans, I received a call from John Pope, a reporter at the *Times-Picayune*. He was writing a story on the surge of out-of-state students applying to colleges in New Orleans because they wanted to be involved in the recovery, and Loyola University had given him my name as a recently accepted student. Pope asked me how I felt about the prospect of returning to New Orleans. I told him I was relieved.[60]

A few days later a copy of the *Times-Picayune* arrived in the mail. Much to my surprise the article, including my photo, was on the front page. It was an unusual welcome to the city and one that signaled a dynamic I would soon spend years trying to navigate. Often white out-of-state volunteers were praised over New Orleanians, specifically Black New Orleanians, who had been there all along doing the same work, and much more. I struggled with the contradiction of needing to physically be in New Orleans to help with the recovery while trying not to contribute to the gentrification of the city as it became whiter, younger, and wealthier. It was a dynamic that challenged me to be constantly aware of power structures in the recovery and the space I occupied within it.

The Loyola University Community Action Program was my first stop when I arrived in New Orleans. It was the student group that did service and advocacy work on campus and in the community. Just a few years post-Katrina, nearly every project we did was in some way related to the recovery. We met once a week in the basement of the student center. There was a room with a couple of couches, no windows, and the faint smell of mold. It was perfect. We packed in around the coffee table, half of us on couches and the other half on the floor, outlining the injustices we saw in the city and devising solutions.

From this little room in the moldy, dimly lit basement, we scheduled thousands of volunteer hours not only from Loyola students, but other universities in the city too. In this room we stored our PB&J fixings for weekend rebuilds, kept crates full of work boots and gloves, and would eventually coordinate our response to the BP Oil disaster along the coast. Around these tables we organized, plotted, did our homework, forged friendships, and pulled all-nighters.

It was a big city and we were a small group, but we tried our best. Time, over a decade of it now, reveals much of the work we did to be small: when an organization needed extra hands, we galvanized the student body into an army of volunteers; when they needed money, we sold crawfish and beer in the campus courtyard; when they needed someone to listen, we invited them to speak; when they needed attention, we used our name to get them media coverage; when they needed support, we protested alongside them. We tried to be creative in leveraging our university's resources for the benefit of the city, doing what we could with the resources we had, but we knew New Orleanians deserved more and better.

The first leadership position I took on was helping to organize groups of volunteers from Loyola to go rebuild houses throughout the city. What New Orleanians deserved and what they got always stood in particularly stark contrast for me on "rebuild days." At 7:00 a.m. on Saturday mornings a group of us—a mix of regular volunteers and first-timers—would meet up in the room in the basement. After calling the few stragglers who had forgotten to set their alarms, consulting one another about the severity of our hangovers, and gathering our lunches, we would head out the door and pile into one of the university's fifteen-passenger vans.

When the flood came, local nonprofits who were working to address these needs and others throughout the city sprang to action. Pre-Katrina, the city had a robust nonprofit sector entrenched in the city as they battled numerous social issues. Prior to Katrina, 28 percent of New Orleanians lived in poverty[61] and 44 percent of adults were reading at the lowest functional level,[62] among widespread issues related to education, transportation, and food access.

Every social issue they had been working to address pre-Katrina was amplified and their ability to meet those needs would determine if entire swaths of the city could return, so they expanded efforts within their areas of expertise. Nonprofits that had worked in the education sector before the flood tried to salvage what was left of the public school system. Groups that had done housing advocacy work ramped up their efforts as the city faced an unprecedented housing shortage and elected officials who were pushing for the demolition of unflooded public housing units. On a good day, many nonprofits struggle to keep their doors open and programs funded. These organizations, particularly the small, local groups in New Orleans, were embarking on a momentous undertaking.

Many of these groups did not have any particular disaster expertise, but they understood they were strategically placed to shape the city's future. They had not planned on doing disaster recovery work but they suddenly had become this recovery's driving force. At the same time many of these groups were themselves affected by the flood. Many groups had lost their offices, equipment, supplies, records, and all the other tools needed to do their work.[63] Their staff and volunteers had also been personally affected and some were even displaced around the country. This didn't stop them. They did what they could to get back to the city. In their offices, they quickly tore down the Sheetrock, treated for mold, and then got to work helping everyone else. In the years after Katrina, if you didn't know better, you'd have thought walls made of exposed studs and creative wiring was the latest office style. Really, gutting their offices was as far as many of them got until there was the time and money to do more.

Despite these efforts, the city had more needs than these existing groups could address on their own. Many new organizations formed in the wake of the flooding, including those which focused specifically on helping homeowners rebuild their houses.[64] These rebuilding organizations became the primary support to many homeowners who had to recover in the absence of government help.

Regardless of which rebuilding organization we volunteered with on Saturdays, our first stop was always their headquarters. In the shell of an office, a volunteer coordinator would give a standard monologue about their organization's origin. They followed a similar pattern: the founder of the group had come to

New Orleans after the storm, seen how much help was needed, gotten together with other like-minded volunteers and New Orleanians, found some funding, picked a street, and started building.

The government failure to lead recovery loomed large over these talks.[65] By this point it had been a few years since Katrina and new students coming to New Orleans were increasingly removed from the flood. I would watch these new volunteers as they listened to these talks and began to recognize what became a familiar look of sudden comprehension and horror as the depth of injustice came into view. I could see the same frustration and anger that I had felt on my first trip to New Orleans. Over and over we heard FEMA had abandoned the city and the federal government had done little to help and had even, for some, made the recovery more difficult.

I had assumed, as many others likely did, that government provides plentiful aid when disaster strikes. As we helped rebuild houses, we talked to homeowners and the employees of the nonprofits who were helping them, about the recovery process. They shared one horror story after another of being stuck in an endless loop of red tape. Through our dust masks and over lunch breaks, homeowners explained that FEMA does not just show up with a check. Survivors had to work their way through a series of complex program requirements before they could be eligible for aid. Even once they did get some help few expected it would be enough to rebuild, let alone make them whole.

On rebuilding days, homeowners would sometimes pull out their "Katrina Binders." When they realized the complexity of the recovery process they would need to navigate, they bought a massive three-ring binder. Over the course of many months, and years, they filled their binder with paperwork from insurance companies, FEMA, banks, contractors, and lawyers. The binders changed color over the years as they were dragged to meetings and became covered in construction dust and specks of paint. The rings would eventually break as they became too full to close. If you asked nicely, homeowners would proudly show you the organizational system they had designed to keep all their paperwork straight.

I always thought the Katrina Binders were a product of this one particular failed recovery. Years later, when I started going to other disasters around the country, I noticed these binders were everywhere. I have seen them tucked under piles of construction supplies on front porches and have seen survivors literally cling to them in the middle of community meetings across the country. These binders are a lifeline that hold the hope of getting the financial support needed to rebuild their lives. Now I understand, these weren't "Katrina Binders," they were "Recovery Binders."

Nothing about the recovery in New Orleans seemed "normal" to me. I would hear politicians talk about the billions of dollars going to the city, but I just couldn't understand where the money was going. It certainly wasn't making its way to the homeowners we were helping. When some money did show up, it was never enough to make a dent in the work that needed to be done. I was so sure that something about the recovery in New Orleans was going wrong. That this wasn't what usually happened after a disaster.

It turns out, that just like the binders, the experience of survivors post-Katrina wasn't actually that different from what many people go through post-disaster. The research I later read supported what I had heard and saw myself in New Orleans.[66] When a disaster happens, we use our own resources first to meet our needs. We use our savings, we run up the credit cards, and try to manage on our own. Unless you're a Kardashian you probably can't afford to rebuild your entire house out-of-pocket with no notice. In fact, this is impossible for *most* Americans.

In 2015, a study found that 46 percent of Americans said they cannot afford a \$400 emergency.[67] They were talking about everyday emergencies like an unexpected car repair or unexpected ER visit. As concerning as it is that so few people are able to afford these relatively standard surprise bills, in the context of catastrophic flooding this statistic has even greater implications. Four hundred dollars isn't even enough money for some families to afford a week-long evacuation, or cover a month's rent at a temporary apartment, let alone cover the cost of rebuilding an entire life.

You may not have thousands of dollars sitting in savings, but that's what insurance is for, right? Well, this gets complicated too. Some types of hazards, including floods, are not included in standard homeowner's insurance. Flood insurance must be purchased separately through the National Flood Insurance Program (NFIP).[68] I've been instructed to keep the NFIP talk to a minimum in this book because it's "boring" so here's the quick breakdown:

In the middle of the last century, private insurance companies stopped covering flooding because it was deemed an "uninsurable risk" (read: it was a huge money loser). Of course, flooding was still rampant across the country, so the government had to step in by creating a federally backed flood insurance program—the NFIP. For many years, homeowners were able to buy flood insurance through the NFIP and the program paid for itself. But in 2005 there were so many claims across the Southern states that the program went into debt, and FEMA had to borrow \$16 billion from the treasury to cover all of the payouts. There was no real plan for repayment to the treasury and so, for the past fifteen years, the NFIP has carried this Katrina debt (with a few billion extra added after Sandy).[69] Many perceive this debt as evidence that the NFIP is an ineffective program. Actually, the debt is evidence that the program did exactly what it was designed to do and that a federally backed flood insurance program is exactly what the public needs (assuming your goal is to get people money to be able to rebuild post-disaster). There are a number of major issues with the NFIP like the accuracy of flood maps, incentive structures, and premium costs. Reform is needed, but the existence of the program is a lifeline for homeowners across the country who need help recovering from flooding.

Despite all this, as of 2018 there are only around five million flood insurance policies, suggesting that as a country, we are dramatically underinsured.[70] It's not only flooding. Almost 90 percent of California homeowners don't have earthquake insurance.[71] These policy numbers are so low for a few reasons, including confusion about what homeowner's policies cover, debates over flood maps, and the cost of premiums. In New Orleans the lack of flood insurance took a particularly insidious turn as many homeowners were told they did not need to have flood insurance—because they were protected by the federal levees.[72]

Even when homeowners have insurance, they may not receive the payouts they expect. Not only do survivors have to wait for home inspections to be completed, but insurance companies sometimes deny that the damage was caused during the disaster, so homeowners either get far less than they expect and/or they have to go to court.

With a quarter of the city living in poverty and with little insurance coverage, there was no way New Orleanians could rebuild on their own. So survivors did what we all do when we need help—they turned to family and friends. The nature of the catastrophe, however, meant that these social networks were strained. Before the storm, nearly eighty percent of people living in New Orleans had been born in Louisiana, giving it the highest rate of nativity of any major US city.[73] This meant that for many, nearly their entire social networks were in Louisiana and virtually everyone in them had also been affected by the storm. Even when people wanted to help their neighbors, friends, and family, there was only so much they could do as they were themselves going through recovery.

The next place New Orleanians turned to for help was government. FEMA had failed them during the response but maybe they could redeem themselves in the recovery (though I don't think anyone was holding their breath). Enter: the Robert T. Stafford Disaster Relief and Emergency Assistance Act, the cornerstone of America's emergency management policy. The Stafford Act outlines the various declaration authorities of the president and the relief programs available to communities post-disaster.[74]

Every day, across the country, people have their homes and property damaged from all manner of storms, flooding, fire, and more. Yet most of them never receive any federal assistance for their damages. Instead, federal aid is given only when the cumulative damages are severe enough to overwhelm local and state resources. There are a series of qualifications you must meet in order to be eligible for individual assistance through FEMA.

First your damage must be from a disaster for which your state (and county!) has received a Presidential Disaster Declaration (PDD). When a disaster happens, the governor of the affected state requests a declaration with evidence that the impacts and needs are significant enough that federal resources are required. With FEMA's blessing, the president decides whether or not to sign the declaration. Only around 75 percent of disasters that are requested are actually granted a declaration by the president, which underscores that it is only the very worst disasters that receive federal help.[75] Assuming the president signs the declaration, various forms of federal assistance are released.

Louisiana easily received a major disaster declaration for Katrina and every parish in Louisiana was approved for various programs. With this formality secured, New Orleanians could begin applying for assistance. Homeowners loved to tell this part of their recovery story to new volunteers because it comes with an obvious shock factor. As part of this application process survivors needed to provide paperwork that proved they owned their home—paperwork that, for many, had been destroyed during the flood.

A further complication was that many never had this paperwork to begin with. As many as 45,000 people, primarily Black homeowners in the Lower Ninth Ward, owned heirs property. This meant their land and homes were passed down from older generations, not through the formal legal system. Without formal paperwork, they were thrown into an expensive legal mess to prove ownership before they could apply for aid. Some estimate that as much as $165 million worth of recovery aid was withheld because homeowners could not produce this paperwork.[76] Twelve years later many Puerto Ricans found themselves in a similar situation after Hurricane Maria.

Assuming these issues were addressed, homeowners then had to demonstrate a need for aid. To prove their need, inspectors view the damage, identify its cause, and assess the amount for which the homeowner is eligible.[77] When "FEMA checks" do arrive post-disaster, it can be a disappointment. The exact amount varies year to year, but as of 2019, the most someone can receive from FEMA individual assistance is about $35,500. Importantly, though, most survivors do not get anywhere near that much money. For example, after Sandy, survivors received an average of $8,000, and after the Joplin tornado, homeowners received an average of $5,7000. Katrina survivors fell in the middle with an average of $7,000.[78] This doesn't come close to meeting people's needs and can make it feel like it isn't even worth the effort. It's also especially insulting when the catastrophe you're trying to recover from was largely caused by the actions of the federal government.

This money also doesn't do much to help cover the costs that survivors incur during the months, and sometimes years, that they spend waiting for aid. If your home is so damaged it's unlivable, then you need to find temporary housing. You may start out sleeping on friends' couches, but eventually you need something more permanent. Finding temporary housing can be difficult post-disaster as there is an increase in demand at a time when housing stock has likely been damaged. This means survivors either have to pay expensive housing prices or they have to move farther away from their neighborhoods.

After Katrina, the housing stock was so severely damaged that even finding any apartment (let alone an affordable apartment) as far as an hour north in Baton Rouge was difficult (enter: the FEMA trailers). Post-disaster survivors can be further displaced, which complicates the logistics of rebuilding. Not only are you working on your home, you're also adding in hours of commuting, and that's before you even take into consideration where your job is located (assuming you still have one), where your kids are going to school (assuming they are able to), and all the other things you need to deal with on a daily basis. You don't just need money to rebuild your house and replace your flooded car: you need the money to rent a temporary apartment, you need money to buy your kid's shoes, you need money to put food on the table, you need money for gas and car repairs. You need money to live.

There are a number of other federally funded assistance programs that survivors may be able to turn to for help. The Disaster Supplemental Nutrition Assistance Program (D-SNAP) works to provide food assistance to affected households[79] and the Disaster Unemployment Assistance program (DUA) is available for survivors who have become unemployed because of the disaster. These are important programs, but after Katrina, they didn't go far enough. The DUA benefits in Louisiana provided less than $100 a week.[80] Further, survivors may find these programs difficult to navigate. The same barriers that people face in accessing these types of government assistance on a good day are amplified post-disaster.

At the risk of stating the obvious—this financial relief often simply isn't enough. Outside of FEMA, survivors are often encouraged to apply for up to $240,000 in low-interest loans through the Small Business Administration (SBA) disaster program. Like with individual assistance, receiving an SBA loan is not a guarantee. Between 2001 and 2018, less than 42 percent of people who have applied for post-disaster loans have been approved, leaving over 860,000 survivors with loan denials.[81] In 2005, 55 percent of applicants (both homeowners and business owners) were turned down, and only 60 percent of the loans that were approved actually made it to survivors.[82] There were widespread reports that SBA manufactured arbitrary deadlines, lost paperwork, and further complicated the application process for survivors, leading some to abandon the program.

The way the SBA disaster loans are approved is a good example of how systemic racism is built into US recovery programs. Loans are given out largely based on the basis of the applicant's credit score, which, for many Black New Orleanians, was affected by the racial wealth gap created by decades of systemic housing, education, and policing discrimination.[83] Across the country, SBA loan applications from people living in majority white areas are nearly twice as likely to be approved. An analysis following Hurricane Matthew found that in a majority Black area of Jacksonville, Florida, the SBA approved 26 percent of applicants compared to 84 percent approved in the majority white community of Ponte Vedra Beach.[84]

Years after Katrina, the recovery process was all-consuming, which helps explains why survivors call recovery "the second disaster." Although both government and nonprofits were doing things to contribute to the recovery, the end result was that many people were not back in their homes, schools were not reopened, businesses remained shut down, and the streets were falling apart for a long time. In New Orleans, three months after the storm, there were still 4,500 people living in temporary shelters, not even temporary housing. Five months later, 85 percent of schools, two-thirds of hospitals, and most food services and transportation routes remained closed. It took years for trash pickup to fully be restored, streetcar lines rebuilt, postal service resumed, and for the hand-painted street signs that had been nailed to telephone poles to be replaced with regular, city-made signs. In 2008, three years after the flood, I personally donned a hazmat suit and entered a house left mostly untouched since 2005. *Three years* it sat in the remnants of toxic floodwaters and Southern humidity transforming mold and mud into a stalactite cave.

This complex set of programs regularly fails to meet survivors' needs, but it was particularly obvious in New Orleans post-Katrina, where the need was especially dire and widespread.[85] These programs especially were not working for Black Americans and other marginalized groups.[86] As this recovery process dragged out over months and years, New Orleanians continued to grow more frustrated with the state and federal governments, and with FEMA especially. They had experienced a traumatic event, made worse because of government, and in the aftermath, when they turned to the federal government to help, they were met with a check that did not even cover their most pressing needs. They couldn't hit Pause on their lives for years to try and figure out these complex and ever-changing programs. The entire process was exhausting in its absurdity.

Many government officials did, to their credit, see that this usual suite of disaster recovery programs would never add up to enough money for many, if not most, New Orleanians to rebuild. It took a while to figure out what to do about it as the state argued with the White House over the best course of action. Finally, seven months after the storm, a plan was sketched out on a paper tablecloth over lunch.

Don Powell, a Texas banker who had been appointed czar of

the Gulf Coast recovery by the Bush administration, and Sean Reilly, a board member for the newly formed Louisiana Recovery Authority, met for lunch one day and made a plan to ask Congress for $4.2 billion to help homeowners return and rebuild.[87] Their plan was for the money to be given to the state of Louisiana and then dispersed among Louisiana homeowners. Congress agreed and the money was used to fund the Road Home program, which was created a full year after the flood.[88]

The Road Home program was plagued with problems from the start. The state of Louisiana oversaw the process, but paid nearly a billion dollars to two private companies to facilitate the program. It took two years for the program to be set up, which left homeowners in continued limbo. Once the program began dispersing money, it became clear that there were significant disparities in how much homeowners were receiving from one neighborhood to the next.

Part of the formula used to determine how much a homeowner would receive was based on the home's pre-Katrina value, rather than the cost of actually rebuilding. This meant that homeowners in Black neighborhoods like the Lower Ninth Ward were receiving less than homeowners in white neighborhoods like Lakeview, even when their home had comparable damage and costs to rebuild. In 2010, a racial discrimination lawsuit was filed against the Department of Housing and Urban Development claiming that the Road Home program was systematically disenfranchising homeowners in poor neighborhoods.[89] The lawsuit was settled with a $62 million payout to thirteen hundred Louisiana homeowners.[90]

The problems didn't stop there. In 2013, around fifty-six thousand homeowners were sent notices informing them Road Home had *overpaid* them and that they would need to reimburse the federal government.[91] We aren't talking overpaid by a few hundred dollars; some homeowners received letters saying they owed upward of $30,000. By the time people received these notices, they had, of course, already spent the money rebuilding their homes (if you're keeping count, we're on year eight of the recovery). To be clear, these were not people who committed fraud; those very rare instances were handled separately. These were homeowners who followed the rules and by no fault of their own were "overpaid," so New Orleanians headed back to court. It took nearly a decade for all the Road Home funds to be dispersed and it ended up costing more than twice the original $4 billion.

Are homeowners expected to wait years for a check? What do they do in the interim? Pay for temporary housing for a decade? Sell their generational property for nothing? Live in a half-built house or in toxic FEMA trailers?[92] This is the reality for many disaster survivors across the country. Perhaps we could do better than a back of the napkin, or in this case, back of the tablecloth, calculation when designing recovery programs.

It is against the backdrop of this absurd patchwork government-led recovery process that the informal, do-it-yourself approach to recovery emerged in the city. Those bare-stud offices of little nonprofits were not just full of "do-gooders," they were the hubs of New Orleans's entire recovery. And those volunteers weren't just helping out—they were rebuilding the whole damn city.

Back in these nonprofit offices, on Saturday rebuilds, the volunteer coordinator answered final questions from the new volunteers. Then we would be given our site location for the day, a printed out MapQuest page with hand-scrawled instructions in the margin that read, like a treasure map, "three lots away from the crooked tree, facing the river," and head out. Street signs had been written out by hand and nailed to telephone poles, but many homes didn't have house numbers. It was years after the storm and there were still entire streets that were just fields. Most of the houses had been demolished and debris removed, but nothing new had been built in their place.

We would lurch our way over the worn, potholed streets, as we passed people hanging out on front porches and construction workers and volunteers just arriving to start the day's work. On-site we were greeted by stacks of tangled tools, music streaming from a radio propped up on an overturned crate, and the sun already heating up.

We'd meet the "site leader," usually a volunteer who was in town working with the rebuilding organization for a few months to a year. They all reminded me of the guy in the white cowboy hat I had met on my first trip. We would gather around and the site leader would recap what the volunteers had accomplished the day before and then confidently outline the tasks we would need to accomplish by our 4:00 p.m. cutoff. The owners of the property were often around as well. They'd welcome us to their home, share their story, and help to keep track of the overall work needing to be done.

It wouldn't be long into the workday before the veneer of expertise would fade and it would be revealed that the site leader knew little more than we did about construction. Within a year of living in the city my expertise surpassed theirs—not something that inspired much confidence. Some days we got lucky and one of the volunteers would happen to have construction experience and they could take over. More often, though, technical expertise remained elusive and we improvised.

Signing up to volunteer in post-Katrina New Orleans meant agreeing to do anything that needed to be done. It didn't matter if you knew how to do it or could do it safely. We dressed in hazmat suits to enter cavernous homes filled with mold. We gasped for air through masks and peered through fogged-over safety glasses in the Southern humidity as we tried to de-mold what remained standing. We tore down walls, laid foundations, hung Sheetrock, mudded, and painted anything that stood still. We took machetes to twelve-foot brush that had grown in place of the not-yet-rebuilt homes. We laid tile and roofing shingles.

We cleared out old gardens, built fences, and fed stray cats who stopped by to watch us work. We fell off ladders and passed out from dehydration. We decided hospital visits for concussions were not necessary (they are). We learned how to get paint out of our hair, and our eyes (the worst). Our wardrobes filled with "rebuilding clothes," covered in paint, bloodstains, dirt, and authentically earned holes.

We didn't just lack expertise; we often did not even have the right tools. We built houses with a single electric saw powered by an extension cord plugged in down the street at a neighbor's house. We used jugs of water, filled off-site, to mix concrete. One Saturday we defied gravity, using hammers to nail Sheetrock to the ceiling. On a phone call home to my father that night, as I lamented about sore arms, he asked, mystified, why we had not just used screw guns? We didn't have any.

Rebuilding the city was as much a part of our regular schedule as going to class and watermelon martinis at Philip's Happy Hour. It was hard work that often wasn't fun. Everything was harder than it should have been. We did not know what we were doing and probably should not have been doing it. Rebuilding an entire city shouldn't have been a DIY project. But we did what we thought needed to be done. Sure, many an imperfectly built home can be found throughout the city, but at least there are homes. To be fair, when I looked around, it was just us—volunteers, this group of ragtag organizations, and New Orleanians piecing the city back together. New Orleanians deserved better than my carpentry skills though. There should have been resources to hire people with training and expertise that could rebuild homes correctly, quickly, safely, and more securely than they had been before the flood. There should have been enough money for screw guns.

Green Dots

ON THE LITTLE black-and-white TV in my parents' kitchen I listened to politicians promise to "build back better" in the aftermath of Katrina. It was the same rah-rah-rah I heard from them after every disaster I'd paid attention to since 9/11. Post-disaster politicians profess their communities are resilient, despite the disaster. They frame the disaster as a battle to be won and refuse to accept defeat as they pit the community against the tragedy. The whole thing is pretty performative but the underlying idea of coming together to conquer the crisis is one that can be inspirational.

Politicians laid it on thick in New Orleans. President Bush stood in a backlit Jackson Square and promised, "this great city will rise again" and we would "build better than what we had before." They promised to spare no resources in the rebuilding and that anyone who wanted to return to the city would be able to. Changes would be made, though, because now there was a window of opportunity to make New Orleans "better." Some of these elected officials also talked about how changes could now be made to the city because it was a *blank slate*. I was in New Orleans only a few days before I realized this didn't add up.

On my first trip to New Orleans my high school classmates and I had spent the week driving around the city, completely lost as we looked for the houses we were supposed to rebuild. There were several near misses as other drivers came barreling out of nowhere, abiding by what seemed like invisible traffic laws. There were no street signs, no houses left to hold address numbers, and if there was a single stop sign left in the city, I didn't see it. Everything had been washed away.

Toward the end of our trip, our group was paired up with a group of local high school students to do rebuilding work together. We had already heard stories from the adults at the homes we worked on, but we were intrigued to hear from people our own age about their experiences. We divided our groups up so we could drive together to our work site. I ended up in the car of one of the local high school students and proceeded to look on apprehensively as he hastily wove in and out of traffic.

He casually chatted with us about the work they had been doing since the storm, seemingly unaware that we were gripping our seat belts, certain we were moments away from crashing into incoming traffic. Eventually it occurred to me he did not need street signs or stoplights. There actually *were* invisible traffic laws. New Orleanians remembered where stop signs had been and what streets to turn down. What felt like anarchy to us outsiders, was muscle memory to the locals.

Disasters do not erase everything. As poet Kalamu ya Salaam said of New Orleans, "It wasn't a blank slate, it was a cemetery."[93] Floodwaters do not destroy the human experiences, memories, history, or sense of community. They do not erase the people who live there and want agency over their futures.

I quickly dispelled the idea that New Orleans was a blank slate, but the goal of building back better stuck with me a bit longer. Building back better is an easy sell. After all, the sentiment is not unreasonable. New Orleans did need to be rebuilt so there *was* an opportunity to improve the community. After all, the state of the city before the flood had facilitated the conditions that made catastrophe possible in the first place. Making changes so the catastrophe is not repeated is the responsible thing to do. New Orleans did need new, stronger levees and flood infrastructure. The electric grid should be rebuilt stronger and homeowners should take future flood risk into account when they rebuild. It is not even possible for a community to be put back together exactly as it was before a disaster, so the idea of rebuilding it better resonates.

As my time in the city lengthened and my geographies expanded, I saw which neighborhoods were looking "better" (Lakeview) and which were not (Tremé). I saw who did get to make decisions (business elites) and who did not (marginalized groups) about what would be rebuilt better. I saw who was paid (private companies with ties to local officials) for these betterment projects and who paid (people prevented from returning home). In New Orleans, building back better meant reinforcing inequality and systems of oppression as well as advancing corporate and elite interest.[94] As writer Margaret Atwood warned, "Better never means better for everyone.... It always means worse for some."

What I hadn't understood when I first arrived in New Orleans is that others heard those same promises—of windows of opportunities and building back better—and understood it to mean something very different. As I became more regularly exposed to these types of recovery speeches, these innocuous phrases became more than just empty promises—I heard them as a threat.

Author Naomi Klein introduced the term "disaster capitalism" in her book *The Shock Doctrine: The Rise of Disaster Capitalism*, published in 2007.[95] The introduction begins with Klein talking to a Katrina evacuee in a Red Cross shelter who was lamenting politicians calling Katrina an opportunity: "It's a goddamned tragedy." Klein goes on to outline her theory of disaster capitalism. In short, elites can use a crisis to implement policies that benefit themselves while the affected community is occupied by their own urgent recoveries. Further, government and elites may instigate the crisis or at least do nothing to stop it. With the public preoccupied, policies and plans that otherwise would have faced opposition are easily implemented, often under the guise of community recovery. This can further a government's existing agenda, which may not mirror the wishes of the community, particularly those marginalized members who are traditionally under- or unrepresented in government.

One architect of this attempted post-Katrina pro-corporate power grab was former Vice President Mike Pence. At the time, Pence was the chairman of the Republican Study Committee. Within two weeks of Katrina the group had met at the Heritage Foundation to develop a list of thirty-two "Pro-Free-Market Ideas for Responding to Hurricane Katrina and High Gas Prices," which Klein called, "straight out of the disaster capitalism playbook."[96]

In New Orleans, disaster capitalism wove its way through the recovery process.[97] While most residents were focused on finding a safe place to live, government took advantage of the "window of opportunity" to make extensive changes to the city's social infrastructure. They moved quickly to privatize the city's public school system and bulldozed public housing in New Orleans, much of which had actually escaped flooding. The "opportunity" was to build a privatized school system and push low-income renters, mostly women of color, out of the city.[98] Louisiana representative Richard Baker admitted as much, saying, "we finally cleaned up public housing in New Orleans. We couldn't do it, but God did." (A comment he later tried to walk back after facing criticism.)[99]

The window of opportunity that was opened by Katrina and

the failed government response wasn't being used to create a city that worked better for all New Orleanians. It was an opportunity to implement an agenda for the powerful at the expense of the marginalized. As I learned about disaster capitalism and the insidious ways it is wielded to reshape communities around the world in the favor of elite interests, I came to understand the terms *build back better, windows of opportunity,* and *blank slates* as the language of disaster capitalists. These phrases are meant to manipulate. They are a palatable articulation of disaster capitalism created from the co-opted hope of disaster survivors. It is rhetoric used to persuade the public of neoliberal agendas.

There were no blank slates in New Orleans. The federal government had no intention of building back better for everyone, and this metaphorical window was opened only to elite and corporate interests. I began to reinterpret much of the turmoil I saw in New Orleans not as profound incompetence among institutions and political leaders, but rather as intentional. I also saw others around the country taking notes. An article published in the *Chicago Tribune* on the tenth anniversary of Katrina speculated that if a similar catastrophe befell Chicago they'd be able to privatize their education system too. Instead of Chicago, in the wake of Hurricane Maria the disaster capitalists descended on Puerto Rico.[100]

A reasonable person would assume that rebuilding an entire city is an endeavor that would benefit from having some sort of plan. In the weeks following Katrina, it became clear that neither the city nor the federal government had a rebuilding plan. Not since the 1906 San Francisco earthquake and fire had a recovery of this size been needed in the United States. There was no modern equivalent. But at the time, the federal government had no blueprint to follow. This uncertainty created widespread confusion over the future of New Orleans and left open for negotiation who would get to be a part of the city.

Following the flood, Mayor Nagin created the Bring New Orleans Back Commission (BNOBC) to develop a rebuilding plan. New Orleanians were scattered around the country and while most people wanted to return and rebuild, exactly when and how they would be able to do so was unclear. Even without a plan, basic services, or instruction from the city, many New Orleanians returned and began to rebuild on their own, but without a clear signal from government it is hard for survivors to make informed decisions. Recovery is a dance and in New Orleans no one knew the choreography. They weren't even listening to the same song.

So when the city finally announced they would be releasing their rebuilding plan, there was widespread anticipation. Based on recommendations from the Urban Land Institute, a national coalition of experts, the BNOBC proposed their plan in January 2006.[101]

A map of the plan hit the front page of the *Times-Picayune* and caused immediate outrage across the city. The map showed six ominous green dots over various New Orleans neighborhoods. The plan was a proposal to turn certain parts of the city into green space. This is how the city thought they would build back better—by bulldozing people's homes and turning the land into parks under the guise of creating a "safer" city.

Developing green spaces as a way to minimize flooding is a legitimate mitigation strategy, and implementing mitigation during recovery is appropriate. However, what the green dot map showed was that some property, often belonging to marginalized residents, would be sacrificed. The white, wealthy neighborhoods would be largely unaffected. There was no green dot over Lakeview. There was no green dot over the French Quarter. There was no green dot over the St. Charles mansions uptown. Instead, the green dots loomed over the predominately Black neighborhoods like the Lower Ninth Ward, Gentilly, and New Orleans East.

The green dot map was a version of the "shock therapy" that Klein identified as a tool used by government and elite's post-disaster. The only way to fight it is through community organizing and that's exactly what New Orleanians did. One green dot neighborhood in particular, Broadmoor, led the charge against the proposal. Broadmoor, compared to some of the other green dot neighborhoods, had more resources. They also had effective leadership, among them the now-mayor of New Orleans, LaToya Cantrell. Through the Broadmoor Improvement Association, subcommittees tackled various recovery issues in the neighborhood and ultimately were able to bring so many residents back to town that the neighborhood was seen as "viable" to the city.[102]

The green dot map was never implemented in its entirety, but in the past fifteen years some parts of the city have effectively become green space. For example, in the Lower Ninth Ward there are large swaths of empty overgrown lots that have not been rebuilt. The city may not have explicitly implemented their plan, but they did implement policies and hold back resources from some communities, which had a similar effect. Pre-Katrina there were fourteen thousand people living in the Lower Ninth Ward but five years post-Katrina the number shrank dramatically to three thousand. Best estimates from 2018 put the total number of people living in the Lower Ninth under five thousand, meaning close to two-thirds of residents did not return to the neighborhood.[103]

As with the recovery planning, when federal dollars began to flow into the city, questions were raised about who would benefit. There were concerns over which neighborhoods would receive the most help, but also who would be on the receiving end of the government contracts. The concerns were well-founded. Mayor Ray Nagin, reelected the year after the storm, was found guilty and sentenced to prison time for a number of corruption charges related to taking favors from companies with city contracts post–Katrina.[104]

While some post-disaster corruption is overt, other forms of corruption emerge out of inefficiencies. After the storm, a federal program was developed to give tarps out to homeowners whose roofs had sustained damage. The idea was that installing tarps would prevent further destruction to the home until a new roof could be built. A Pulitzer Prize–winning investigation by the *Times-Picayune* found that the federal government was paying up to $5,000 to have a blue tarp installed on a house, while the workers who were actually installing it made only a fraction.[105] One company received the federal contract and they subcontracted to another company, which then subcontracted to another company, which then subcontracted to another, all the while they each withheld a little bit of money, until the workers who actually did the work earned a pittance. Absent government oversight, this subcontracting scheme would have continued if not for local investigative journalistic efforts. When the federal government cannot effectively facilitate a contract to hand out tarps, it does not bode well for how the rest of recovery will go.

Corruption, inefficiencies, backroom deals, and opportunists with selfish intentions permeated the recovery. They all blended together, so it was impossible to tell where one ended and the other began. It left the city in a perpetual crisis, and me confused. To obscure and disorient is, of course, part of the strategy.

This approach to recovery is designed to appease the needs of the white middle class, provide a safety net for the wealthy, clear the way for elite interests, and for businesses to make a profit by exploiting the marginalized. This is what they mean by better.

Disaster survivors don't just have to survive the actual disaster; they have to survive the recovery too. By the time I lived in New Orleans, the green dot map had largely been put to rest, but the incident became local lore, highlighting how local government sought to take advantage of its most vulnerable communities and failed to execute a just recovery.

My drive home from class at Loyola required a shortcut through the Broadmoor neighborhood. The route was familiar as I drove it twice a day. The threat that the green dot map had once posed seemed a piece of the past. The neighborhood looked to be on the brink of thriving again. On my commute, I had seen over the course of many months the construction of the new neighborhood library, which was slated to also have a café. One day, as I drove home, I saw a sign had been added in front of the new library. The sign, in the shape of a green circle, read in white lettering: The Green Dot Café.

Bootstraps

ON ONE SATURDAY afternoon in 2010, I stood on top of a house in the Lower Ninth Ward laying new shingles. It was one hundred degrees, and I had been up there all morning. Roofing tar had melted into my palms, my knees burned under my jeans, and I barely had enough energy left to lift the nail gun. Over time this kind of work had become both more familiar and more exhausting. Saturday rebuilds were beginning to feel like *Russian Doll* (see also: *Groundhog Day* for us olds). It had been five years since the flood—half a decade.

The construction symphony still rang throughout the neighborhood, but it was quieter now than it had once been. The smell was long gone and the air didn't stick in quite the same way. The urgency, at times near frantic, that had once driven our workdays, was replaced with a more appropriate Southern pace, driven by the heat, limited resources, burnout, and departure of many volunteers. I looked out over the neighborhood at the homes peeking out between empty, overgrown lots. Behind me stood the now-rebuilt flood wall along the Industrial Canal and some of the houses I had worked on my first time in New Orleans. In front of me sat half a decade of the accumulated efforts of New Orleanians, a million volunteers, contract workers, and day laborers, and the entire United States government. So much work had been done but a recovered city, one where Katrina was talked about in the past tense, was still elusive. This was disaster recovery in twenty-first century America, and I was far from impressed.

That afternoon while I was physically in New Orleans, my mind was seventy-five miles offshore. Deepwater Horizon had exploded a month earlier and oil had been gushing out of the ocean floor into the Gulf of Mexico for weeks. Just days before, I had been down in Grand Isle working with community organizers to bring media attention to the needs of coastal residents suffering the physical, emotional, and economic impacts of the disaster. Handmade protest signs lined the road through Grand Isle that were expressive, shall we say, toward BP executives and the Obama administration. The signs' messages were reinforced in the conversations I had with locals. Tensions were high and emotions raw as the disaster dragged on. I checked NOLA.com on my lunch break for any updates on the well. As had become the norm, there was no good news. The oil was still flowing freely and efforts to cap the well a mile below the ocean's surface continued to be unsuccessful.

I had come to understand disaster recovery as an exhausting, unjust, confusing, and slow process, but I had to that point always tried to focus on the movements forward. That afternoon I felt defeated, as I would for many months after. The BP disaster felt like throwing Louisiana back to the starting line. We had not even finished cleaning the toxic flood water out of the streets of New Orleans and now there were millions of gallons of oil to clean up along the shore.

I wanted to stand on the roof and scream. Was I dehydrated and suffering from heat exhaustion? Yes. Was my anger totally justified? Also yes. BP killed the part of me that still clung to the belief that Katrina was an accident, an outlier. The magnitude of injustices, so physically and temporally close to one another, forced me to confront what now seems obvious—disasters are not freak accidents, they are the inevitable product of the decisions we, or some people, make. They are not once-in-a-lifetime events. They happen every day.

When I moved to New Orleans, I expected to stay until the recovery ended. But by the time I neared graduation, seven years after Katrina and two years after BP, I could not see the finish line. Moreover, I was not even sure anymore that there was a finish line. With another decade-long recovery about to begin along the coast, I wondered if recovery was even achievable. In a place like southeast Louisiana where disaster persists, I struggled to see where one recovery ended and the next began. I thought about how the concept of normal ceases to exist for communities that are trapped in cycles of recovery. The promise had always been that with one nail at a time we would rebuild the city. The time I was spending rebuilding was impactful for the people who lived in those houses but weekend volunteer rebuilds felt like putting a bandage on a gushing artery.

I had always had this feeling that one day the cavalry would show up. I never fully realized we were the cavalry. We were doing good work. There were New Orleanians who were back in their homes because of the work we did, but there were also thousands of New Orleanians who were not. We were not enough no matter how much good work we did. I was, for the first time, experiencing the inherent heartbreak of disaster work—no matter how much good you do, it won't ever be enough to help everyone who needs it.

I assumed the disproportionate recoveries and inefficiencies unfolding in New Orleans were not normal. I thought something was going wrong in New Orleans, and it seemed others did too. It felt like someone, somewhere had made a mistake.

Many placed the blame fully at the feet of a few politicians and the federal bureaucracy, but it was more complicated than that.

At Loyola my understanding of the recovery evolved one book and one class at a time until I learned a more comprehensive narrative; one that accounted for how a history of political corruption and ineffective leadership collided with slavery and Jim Crow to create generational poverty. I studied the history of New Orleans, the urban geography of the city, the state's history with disasters, and even the ecology of the Mississippi River Delta, until I could see the direct line from Jean-Baptiste Le Moyne de Bienville's settlement of the city in 1718 to the 1927 Mississippi River flood, through the civil rights movement, to the levees breaking in 2005.

When the BP recovery began, I saw the same problems rising up outside the city: the government was scrambling, there was conflicting information, needs were not being met, and it was the nonprofits and volunteers that were standing side by side with those affected. Access to resources was delivered along racial and class lines, politicians bowed to corporate interests, those affected were dependent on holding the national media's attention to receive help—all while nonprofits triaged the most pressing needs. I considered that perhaps this tumultuous approach to recovery, and emergency management more broadly, wasn't just a Louisiana problem.

A year later in May 2011, a mile-wide EF-5 tornado ripped through Joplin, Missouri, killing 161 people and injuring over 1000.[106] It took the title of the country's most expensive tornado with nearly $3 billion in losses. Over 500 businesses were damaged and 7,500 homes were damaged or destroyed along with the high school, several elementary schools, and the hospital. A few months later I stood in a pile of dirt that had been someone's kitchen. Around me, every direction afforded the same view: empty dirt lots where the Missouri suburbs had once stood.

The few tree stumps that had withstood the tornado sat waiting to be de-rooted by volunteers. The ground was blanketed with lifetimes of material objects that had been blown all over town. There were damp magazines, bent silverware, and assorted crushed Matchbox cars that were swept up into wheelbarrows.

The Joplin tornado was devastating, but it did not, in either death toll or damage, compare to Katrina. Yet, despite these clear differences, the recovery process looked familiar to me. Like in New Orleans, the only people we encountered out working were volunteers who once again seemed to be the only cavalry around. Some elements were more organized, and, by the very nature of there being fewer impacts, the process was moving more quickly, but the challenges at the heart of New Orleans's recovery had moved north to Missouri: bureaucratic hurdles, misinformation, and inadequate financial assistance to survivors. We were in a different state, dealing with a different hazard, on an entirely different scale, but the inefficiencies were the same.

It was in Joplin, among the debris of people's lives, that I finally had something with which to compare my experiences in Louisiana. I realized that slow disaster recovery was not a Louisiana problem, but a national problem. Acknowledging the regularity of disaster recoveries shifted my perspective. If disasters happened so frequently and communities had to recover so often, why were we so bad at it? And why did people not seem to know disaster survivors were being put through this? On a church floor in Joplin, where we slept, my body was exhausted but my mind raced as I wondered how we could continue to allow people to go through this turmoil again and again, without doing anything to make it better.

I could see no end to the houses that needed to be rebuilt in the Lower Ninth Ward, in New Orleans East, in Tremé, in the Lower Garden District. I could see no end to the people along the coast who were working to find new livelihoods in the wake of the BP disaster. I could see no end to the families in small towns across the country who were trying to recover from their own disasters. Watching the Katrina recovery unfold was one thing—I had signed up for that—but when I realized the problem was not confined to rebuilding New Orleans, but rather every community affected by crisis, I became overwhelmed. No matter how many houses we fixed in New Orleans there would always be one more house, in one more neighborhood, in one more city, that needed to be rebuilt. And the disasters were happening faster than we could rebuild. I couldn't be in Joplin and New Orleans at the same time. How can you choose one community over another when they both need and deserve help?

Back in New Orleans my confusion and frustration came out, possibly a bit neurotically, in an essay I wrote for a class on Katrina recovery that I took at Loyola. My professor, who had studied disasters at the University of Delaware's Disaster Research Center, must have suspected I was only a few more Saturday rebuilding trips away from a complete meltdown, because she sat me down on a bench in the quad after class and very nicely told me to go straight to graduate school. I needed a language to describe what I was seeing, and graduate school was where I could find it. I agreed. I felt like we were always playing catch-up. We were always just reacting to the most recent disaster. I thought that if I could learn more about how we should manage disasters, I could then help implement change for the communities that would one day face a catastrophe of their own. I thought we could be more proactive. I didn't want anyone else to have to go through what New Orleans had been through.

A few months later I selected the emergency management program at North Dakota State University and moved to Fargo where I immediately experienced culture shock. No two cities in the United States are more diametrically opposed to one another than New Orleans and Fargo. I felt like I had moved to a different country. The people, the food, the culture, the music, the weather, and the landscape were each other's exact opposites. Walking around downtown Fargo felt like walking around Disney World. The sidewalks did not have cracks, there were no potholes, the streets were clear of litter. The storefronts were perfectly curated. City workers watered the hanging plants on every lamp post. It was pristine.

It was not until I moved to Fargo that I realized the extent to which I had assimilated to driving on those cavernous roads, being rerouted daily for sewage system projects, boil-water advisories, and rushing out to move my car every time it rained because the city drains and pumps were not fully operational. These small adaptations had become part of my daily life in New Orleans. Even when I wasn't conscious of it, everything I did and everywhere I went was influenced in some way by the recovery. Often we talk about recovery as something that happens to a place—but really, recovery becomes the place. It becomes inextricably intertwined with the community, weaving its way into the fabric of daily life.

As I started reading the disaster research, I learned that not only can the trauma of living through a disaster contribute to mental health impacts among survivors, but the stress of the recovery process itself can as well. Psychologists have found an increase in stress among individuals going through recovery, can lead to negative health consequences. Stress can worsen preexisting and chronic conditions[107] and can contribute to an increase in suicide rates years after the disaster.[108] These deaths, and others indirectly caused by the disaster, often go unrecognized in official death tolls.

There is also often an increase in domestic violence post-disaster. The research on why this happens is still in an exploratory stage.[109] Initial evidence suggests that abusers may exploit the post-disaster landscape of uncertainty by providing housing and other resources to domestic violence survivors who had previously left abusive relationships. Further, relationships in which one partner had controlling tendencies pre-disaster may escalate to physical violence post-disaster.

These issues—chronic health conditions, suicide, and domestic violence—all come at a time when the community is likely least able to address them. For example, women's shelters that are usually available to domestic violence survivors may have been affected by the disaster themselves. This happens at the exact moment there is an increase in demand for their help.[110] Similarly, the increased demand for mental health services may coincide with those services becoming more difficult to access.

Disasters, of course, do not just create new problems; they exacerbate old ones. The city of New Orleans was already battling a mental health crisis before Katrina. Then the flood came, and compounded food insecurity, unemployment, and homelessness. The local healthcare system fell apart as Charity Hospital, the primary provider of psychiatric care, was damaged during the storm and shut down despite city-wide opposition.[111] Within the city itself the only option for mental health services was a temporary stay at the Louisiana State University Medical Center, which was exceptionally understaffed and not equipped to provide long-term care.

Unless you were incarcerated. For years after the storm, the Orleans Parish prison remained the largest psychiatric facility in the city.[112] Anyone else seeking help had to travel outside the city, a difficult, if not impossible, task for those without access to transportation or financial resources. In the years after Katrina, suicide rates in the city tripled.[113] The official Katrina death toll does not include these people or those who died months after the storm from stress and chronic illnesses for which they were kept from accessing healthcare to address.

All the while the media moved on, volunteer numbers dwindled, government sent survivors in circles, and double-digit anniversaries approached. As I read this research, I was struck by the fact that we know this happens—disaster after disaster—and yet relatively little is being done to address these issues in recovery. As I continued through graduate school, I became even more horrified by how much research there is about disasters and how little of it has made its way into practice. A century of disaster research is just sitting here, and yet changes in the field, based on this work, are rare.

It so happened that the first course I took at NDSU was Disaster Recovery. I sat in class the first day, awestruck, as our professor casually mapped out an entire framework for understanding why and how we manage disasters. Everything I'd spent years confused about, she explained in forty-five minutes.

I had learned the history of Louisiana in my classes at Loyola so I thought I had a good understanding of Katrina's context. But in my grad school classes I learned the history of the US approach to managing disasters and realized I had been missing a key piece of the puzzle. The recovery in New Orleans was not a *failure* of the system as I had thought, but rather it was the *intention* of the system.

Prior to the 1930s, the federal government was rarely involved with helping communities affected by disaster. When a disaster was big enough and those affected had a powerful advocate in Congress, aid in some form may have been allocated, but it was only done on a case by case basis.[114] The traditional story of recovery relies heavily on a narrative that Americans "pulled themselves up by their bootstraps." What those histories conveniently leave out is the experience of marginalized groups, not only in how they themselves were affected by disaster, but also in how their forced or meagerly paid labor was used in the recovery.

Devastating events, including the Dust Bowl and Great Depression, on the heels of the 1927 Mississippi River flood, led to a shift in the federal approach to managing disasters. Years of crises affecting the entire country revealed the limitations of self-sufficiency. Washington began to face growing pressure for federal intervention and involvement through disaster relief programs. Over the next fifty years, disaster policy was passively cobbled together as Congress parachuted in to sporadically help after some disasters. Although federal involvement in disaster recovery has steadily increased over time, the federal government still remains limited in its involvement.

Specifically, the US uses a "limited intervention model" of rebuilding.[115] After disasters, even catastrophes like Katrina, the federal government is not attempting to fully provide for survivors. They intentionally limit the help they give with the expectation that individuals will pay for their own recovery using their savings and insurance payouts.

Our approach to recovery is rooted in America's dominant political ideology—that the government should limit its involvement in citizens' lives. The government plans to provide some help for the very worst disasters, but otherwise they leave recovery up to individuals and the private sector. So while politicians get on TV and promise everyone will recover, there is no guarantee, and no attempt by the federal government to facilitate this.

This approach works fine for some people who have minimal damage or have access to substantial financial resources. People who have hazard insurance, savings accounts, and stable incomes may be able to pretty easily reach a place of feeling recovered. People who have the time, ability, and resources to navigate the recovery process can find some help through SBA loans or FEMA's individual assistance programs. But this does not work for everyone.

This limited intervention model also assumes an individualistic approach to recovery. As I watched recovery unfold in New Orleans, I saw that recovery is a social process. Much of recovery is completely outside the control of both individuals and nonprofit organizations. In New Orleans, homeowners were doing exactly what they had been told to do. They had drained their savings, asked friends for help, and applied for the government assistance that they were rightly owed. Yet, years later, many had still not reached a point of anything approximating recovery.

When New Orleanians returned to the city, they faced these daily roadblocks that were largely outside of their control. They were dependent on the decisions of others around them and had little to no control over what they would do, how they would do it, and when. New Orleanians did not get to decide when power would be restored to their neighborhood. They did not get to decide when the post office would start delivering mail again or when the city would start clearing debris and picking up trash. They did not even know if and to what extent the government was going to rebuild the levee system or if their neighborhood would be turned into a park.

Decisions made by others in the community, government agencies, and the private sector were shaping the course of their recovery.[116] These decisions affected the speed of their recovery, often bringing it to a standstill. It was not just about getting their home rebuilt, but making sure the roads were fixed, schools and hospitals were open, and the electricity was restored. New Orleanians had to stage what essentially amounted to sit-ins at city hall to get building permits and at Entergy to get their electricity restored.

We may now have a glossy National Disaster Recovery Framework,[117] but the underlying bootstrap approach remains—complete with the most marginalized people in our country bearing the greatest cost. Disasters illuminate underlying social inequalities, but the system reinforces the very issues that led to the creation of the disaster in the first place. It is a paradox of disasters that the people with the fewest resources are most likely to experience the worst effects of disasters. Years of policies have funneled the economically poor into physically vulnerable areas. A lack of money and institutional support has exacerbated this vulnerability as people are unable to afford mitigation or preparedness efforts like raising their homes to prevent flooding or buying insurance. So when a hurricane comes along, it is these areas with higher physical vulnerability that bear the brunt of the storm and suffer the most severe consequences. While response reveals inequality, recovery reinforces and even deepens it.

Once the immediate crisis is over, survivors find minimal support to help rebuild their lives. For many, their own resources are not enough, and with nonprofits unable to do more and government unwilling to, it is not clear how survivors are expected to recover. They certainly do not have the resources to "build back better" or mitigate future disasters as they rebuild. Then they are chastised when the disaster happens again. It is a ruthless cycle steeped in centuries of racist and classist policies and that is out of line with a proactive approach to minimizing community vulnerability. This system has not worked for many people for a long time. Katrina was just the first time that I happened to witness the problem myself.

Visiting other places across the country that were recovering showed me that although these different places had experienced events of differing scopes, scales, impacts, and types of hazards, their recoveries were unmistakably similar. It was the same story, different disaster—a mantra I now say with such frequency it feels contrived. One community going through this ineffectual process is a failed recovery, but that all communities experience this turmoil indicates the failure is systemic.

What I now understand is that America's recovery system is not broken, it is working exactly as designed. When I learned the history of the US approach to managing disasters I realized there was nothing unique about New Orleans's recovery process other than its scale. The approach taken there was the same one taken after all disasters. Hoping for government benevolence in the face of disaster is enticing, but in a country where the education system, healthcare system, and criminal justice system are all designed to uphold white elitism, why would the emergency management system be any different?.

Commemorative Snow Globes

A FEW MONTHS before the ten-year anniversary of Katrina, I sat on a set of crumbling concrete steps with a rusting wrought iron railing that a decade earlier had led to someone's front door, but now led to an abandoned and empty foundation. The stoop was all that remained of the home that had once stood there and had never been rebuilt. I was there to interview the group of college students who were rebuilding the house across the street while on a spring break immersion trip to the Lower Ninth Ward.

It was part of a larger study I was doing of disaster volunteerism in long-term recovery. People always show up to help in the early days of recovery, but Katrina was unusual in terms of the extent of volunteerism and the length of time it was sustained. In 2006, there were an estimated ten thousand volunteers coming to the city each week to help with recovery. In the decade and a half since the flood, well over a million people have come to help.[118] High school and college students, church groups, corporate volunteers, and groups of friends continued to come to the city to help, alongside New Orleanians, for over a decade. Volunteers tend to lose interest as time progresses,[119] and while the flow of volunteers has slowed in New Orleans, it has yet to completely stop, as of this writing.

Many of these volunteers were like me when I first came to New Orleans in high school. We organized our own trip, spent the majority of our time volunteering directly with local organizations, slept in sleeping bags at a volunteer camp, ate brown-bag lunches, and ventured into the French Quarter on our afternoon off. It was a go with the flow kind of experience. Over time though, these improvised trips transformed into a voluntourism industry.

New Orleans's distinction as a tourist destination was a natural draw for people from around the world. The French Quarter was quickly cleaned up to welcome visitors and the city unveiled a new tourist attraction: Disaster Recovery. Tourists could have the fun of Bourbon Street, while also spending a few days rebuilding. The combination laid the foundation for a scale of disaster voluntourism never seen before, or since, in the United States. For-profit companies offered travel packages for people who wanted to dedicate part of their vacation to rebuilding homes. They arranged airfare, hotels, car rentals, and activities like swamp tours and visits to Mardi Gras World, all while connecting their customers with a nonprofit doing recovery work. The director of one local nonprofit once described these companies to me as "volunteer traffickers" because of the constant stream of volunteers being funneled into the city.

One by one the college-aged volunteers came to sit on the stairs and tell me about their experience. A decade later, I heard them repeat my own experience back to me. They had been presented with the opportunity to come to New Orleans. They had not known much about the storm or levees. They were shocked by the extent of remaining damage they found on their arrival.

They were confused and angered by the lack of government assistance. They had a fun afternoon on Bourbon Street.

If I had been interviewed during my first trip it would have been indistinguishable from theirs. It was startling to see a reflection of myself at sixteen, a decade later. On one hand, I had learned in graduate school that nothing about my experience in New Orleans had been unique. It was the same experience that a million other people had found when they came to help with the recovery efforts.[120] It is an experience that is mirrored in dozens of communities around the world every year. But it was still jarring that the volunteer experience in New Orleans had remained largely unchanged in ten years. You expect, after many years, things will be different or at least that you'll find time has exaggerated your memories. There on those concrete steps, it was all the same. A tourist attraction frozen in time.

It also did not matter that it had been ten years since the levees broke; the volunteers unequivocally described their efforts as recovery work. The organizations that hosted them agreed. These local nonprofits, many of which were the same ones I had worked with years earlier, had certainly slowed down, but they were still actively engaged in the recovery process. I had to agree with the volunteers. There were hundreds of empty lots that had never been rebuilt, many now rewilded, and nearly every block had an active construction site. For every national nonprofit that had departed, there were two local groups left in its place that were still welcoming volunteers regularly from around the world.

Yet, on the ten-year anniversary, there were celebrations. Mayor Mitch Landrieu used the annual State of the City address to announce that New Orleans was "no longer recovering, no longer rebuilding" but rather "creating."[121] The city threw a parade and President Obama came to give a speech.

It was difficult to reconcile the celebrations with what I still saw in some parts of the city. The mayor's claim felt like an erasure of the ongoing work being done by many volunteers and organizations. Worse still, it was an erasure of the New Orleanians who were still not back at home. Certainly, there were some, maybe even many, New Orleanians who found comfort in the mayor signaling the end of a difficult decade. There were people who had recovered, and it was important to celebrate that success. But it felt like those celebrations were void of interrogation as to why not everyone had.

Despite this, I did have a deep emotional connection to the city and the entire performance of the tenth anniversary left me conflicted. I was disgusted over businesses selling commemorative Katrina snow globes. I was gutted listening to survivors' stories again. I was nostalgic for what Rebecca Solnit calls the "paradise built in hell"—the feeling of a community coming together and helping one another. I was relieved to hear the many stories of survivors who had been able to recover.

Mostly, though, I was angry. The anger I had felt during my first trip, at sixteen years old, had both deepened and transformed. I was angry that Katrina had ever even happened. I was angry there were still people who needed help. I was angry that many of the changes that we knew needed to be made to the emergency management system had not been made.

I felt an exhausted frustration about the disaster researchers and other experts who had known how a storm like Katrina would unfold before it ever formed in the Gulf. Too many experts knew the levees could fail. They knew the Ninth Ward would be a field. They knew the French Quarter would come back thriving with tourism. They knew Young Urban Rebuilding Professionals (YURPs), the name for the mostly white twenty- or thirty-somethings who moved to New Orleans during the recovery, would arrive to continue the trend of gentrification.[122] They knew there would be congressional acts, impressive in name only. They knew there would be people who would never return home. They knew FEMA couldn't handle a catastrophe and they knew the White House wouldn't.

Carrying this anger had propelled me from voluntourist, to YURP, and now to disasterologist. I was especially angry because now I knew that I was one of the people who knew all of this was possible and like them, I felt powerless to do anything to stop it from happening again.

By the tenth anniversary no one asked me about the recovery in New Orleans anymore, which was a shame because by then I had found the words to explain the injustices. That's okay. I had gotten good at bringing it up myself. Anyone within my general vicinity was, and is, treated to a lecture on how "community revitalization" was a palatable term for government failure. $160 billion and ten years and they still had not met all of their residents' basic needs. Americans expected a success story, though, so in August 2015 the media delivered with a predictable deluge of articles that tried to rewrite the history of the failed response and declare the city recovered. To me, the ten-year anniversary felt like a cover-up.

The guy in the white cowboy hat was wrong. It wouldn't take a decade for New Orleans to recover, at least not all of New Orleans. It would take much longer. It has been fifteen years, as I write this. People are still rebuilding their homes. The summer of the fourteenth anniversary I spent a day volunteering with a group of emergency management undergrads, building a house in the Ninth Ward. Around us there were plenty of houses that had been rebuilt, but when I looked just at this one house, it was as though time had stood still. I watched the group of eighteen- to twenty-one-year-olds from Missouri put the framing up. They hung off ladders, and I pestered them to put on sunblock and drink water. Neighbors slowed as they drove by, waving out the windows and shouting a thank-you. Fourteen years after the flood and it could as well have been my very first trip there. I kept an eye out for the lady in the purple outfit.

I imagine there is a lady in a purple outfit in every town, standing her ground. She is refusing to be moved. Fighting against inefficiencies, corruption, and racist recovery policies that prevent her from returning to the place she calls home. She often loses, but sometimes she wins.

It's wrong, though, to mistake the lady in the purple outfit as a beacon of hope. Her presence is indicative of a national failure and her strength—a strength she should not need to have. It is not the responsibility of survivors to make you feel hopeful. Their lives are not a story meant to make you feel more grateful for your life. They are not a life lesson for you. They are people—your neighbors who need your help. It is your job to do something, anything, to stack the cards in her favor. Better still, it is your job to prevent the lady in the purple outfit from ever having to be in the position of needing to fight for her life.

PART TWO

Mitigation: Preventing Disaster

*"When I die, the scientists of the future,
they gonna find it all.
They gonna know, once there was a Hushpuppy,
and she lived with her Daddy in the Bathtub."*

–QUVENZHANÉ WALLIS, *BEASTS OF THE SOUTHERN WILD*

THE LITTLE SEASIDE city of Saco sits about one hundred miles north of Boston on the coast of Maine. Driving north on Route 1, you will pass through the center of downtown where, if you turn right down Beach Street, you'll snake your way past family homes, a garden center, a variety store, and an ice cream shop until you arrive at the entrance to Camp Ellis. At the intersection of the Saco River and Atlantic Ocean, Camp Ellis was once a bountiful fishing village. Now the neighborhood is a small-time tourist destination featuring a mix of year-round locals and summer renters.

Camp Ellis is a six-by-seven-block neighborhood, kept from being a perfect grid by the protruding estuary. The houses, mostly summer beach cottages, are weathered and typical of the New England coast. They are separated from the ocean by a strip of sand that grows larger at low tide and a jumbled rock wall built of imposing gray boulders. Sand-covered streets weave their way between the houses with yards full of old beach chairs and washed-up buoys. The roads are narrow and there are no sidewalks, which forces pedestrians to turn to the side as cars pass slowly by, greeting each other with a wave.

At the far end of the neighborhood, a worn parking lot comes to an end at the working waterfront where lobster traps are unloaded onto the wooden dock. In the summer, the driveways fill with visitors from the rest of New England. It is a small neighborhood, in a small town, with people who spend the summer fishing and the winters fussing about snow.

Despite growing up only twenty minutes north in another town at the edge of the sea, I had never spent time in Saco until my parents moved there in 2015. I got to know their new town on my visits home, including long drives down the winding road to Camp Ellis. As a haven in the fast-paced Northeast, the little coastal neighborhood promises to shield out the rest of the world; but as I got to know Camp Ellis I learned something sinister was happening there. Beneath the veneer of idyllic New England life, Camp Ellis is in the final throes of a one-hundred-fifty-year battle for survival. Extinction is likely, if not imminent.

In early January 2018, a blizzard set its sights on the Southern Maine Coast.[123] While my family pulled out the generator, fueled up the snowblower, stocked up on hot chocolate, and settled in to watch the running list of closures grow on Channel 6, I snuck down to the beach. The streets of Camp Ellis were among the first to flood. Outside it smelled like snow and the ocean pulsed as spray flew into the air and the waves slammed against the rock barriers before pushing the tide into the neighborhood.

A well-rehearsed dance unfolds when it starts to flood in Camp Ellis. Barriers are put up to keep out sightseers. Homeowners move their cars down the road to keep them dry, tie down loose objects in their yards, and move their valuables to higher floors. Boards are pulled out of sheds and hung over windows. Sandbags are propped up against doorways. Some residents evacuate, but others stay to manage the water as it approaches. Reporters drive down from Portland and livestream the waves overtopping the riprap seawall.

As the storm passed, city officials came to assess the damage. The city came with backhoes to clear the roads of sand and placed concrete barriers to block off the roads too damaged to use. Residents returned home to take stock of the damage. They checked in on their neighbors and retrieved their belongings that were swept down the street. They shoveled sand off their driveways and front walks and called contractors.

Despite the flooding in Camp Ellis, the storm was far from the worst-case scenario. Across the state thousands lost power, a few streets in downtown Portland flooded, and flights were canceled, but there wasn't enough damage to warrant a Presidential Disaster Declaration. There were no FEMA checks. The national disaster nonprofits did not descend. Donations were not collected. The national media didn't pick up the story. Mainers were on their own to put the pieces back together. Then came the next round.

In March another storm battered the coast of Southern Maine.[124] Again, residents moved their cars to high ground and barricades were returned to the entrance of the neighborhood. Fourteen consecutive high tides flooded and reflooded Camp Ellis and other coastal neighborhoods. Still in the midst of recovering from the January storm, Camp Ellis took a beating. Dozens of homes were damaged. One came precariously close to collapsing into the ocean and others were buried in sand up to their porch railings. The roads flooded, cracking in places under the force of the water. The road closest to the beach, North Ave, disappeared completely under a mountain of sand.

In total, the March storm caused millions in property damage statewide. Compared to the billion-dollar disasters other states have had to contend with, the March storm was a blip, but for a small state like Maine it was significant. York County, home to Saco, reported the worst beach erosion in state history. Saco itself reported $650,000 in damage and the severity of impacts to infrastructure in Camp Ellis was equivalent to the notorious Patriot's Day storm of 2007.[125]

Saco had made it through the January storm on its own, but the recovery costs for the March storm eclipsed what the town could reasonably afford. In combination with neighboring towns, the county had enough damage to request assistance from FEMA throughout the state. In a rare win, a major disaster declaration was signed and over two million dollars put toward public assistance in York County.[126] These winter storms and the damage they caused was nothing new for Camp Ellis. In fact, the neighborhood's flood problem began with a decision made by the US Army Corps of Engineers one hundred-fifty years earlier.

Creating Disaster

EVERY DISASTER YOU have yet to experience in your lifetime has already begun. The threads of risk are spun out over decades, even centuries, until they crescendo into disaster. In Hollywood, disasters happen fast but in real life they are slow, which is why you need to know the long story of how Camp Ellis became Camp Ellis.

Camp Ellis was originally home to the Abenaki. Europeans including Martin Pring, Samuel de Champlain, and John Smith noted the area as they mapped the New England coast in the early 1600s.[127] By 1630 the English Council of Plymouth issued a patent for the land that eventually became Saco, and within a few years almost forty English families lived in the area.[128] Throughout the seventeenth century, the Abenaki people defended the area from European invasion and colonialization in a series of six wars ending with the French and Indian War of 1760.

During the Revolutionary War, Saco and the neighboring town of Biddeford protected the mouth of the river from British invasion. After the Revolution, Saco's population grew steadily, bringing with it economic and land-use development in line with the rest of New England. The coast industrialized and mills were built a few miles upriver in Saco to produce nails, textiles, and lumber. At the turn of the nineteenth century, more than a dozen sawmills produced fifty thousand board feet of lumber daily,[129] which was shipped down the Saco River to go build the rest of the country. As one of the first areas settled by Europeans in what would become the state of Maine, the port in Saco became a centerpiece of the region's economy.

Outside of town, at the entrance to the Saco River, a neighborhood, soon to be named Camp Ellis, began to grow. Prior to the 1800s, Reverend Ellis owned the only house in the area, but eventually more homes and a church were built among the dunes. In the winter the river froze, rendering it impassible, so ships had to dock at the Saco Ferry, south of the mills and just north of Camp Ellis. The activity at the winter port drove further growth in Camp Ellis, including the building of more family homes and the opening of a few stores.

In the late 1800s, the Army Corps of Engineers were on a nation-wide mission to tame nature, and the Saco River was on their list. In nonwinter months when low tide arrived, the entrance to the river became blocked by a sandbar. Ships had to wait for high tide in order to travel up to the port. If, however, ships could navigate the river regardless of the timing of the tide, they would be able to double their profits.

At the behest of Congress, the Corps proposed building a jetty at the mouth to stabilize the river and allow ships unabated access to the mills upriver at Saco-Biddeford.[130] The jetty, really just a long pile of rocks, would stretch from the tip of Camp Ellis out into the churning Atlantic. Construction began in 1828 and by 1867 the river was clear for passing ships.[131] The jetty, at first, was a success, and businesses boomed. Residents, however, soon began to notice the beach in Camp Ellis was eroding, while the beach across the river in Biddeford was growing.

The jetty had changed the movement of sand.[132] Sediment from the Saco River had previously been deposited along the shore of Camp Ellis, building out the land, but when the Corps built the jetty, the nearly twenty thousand cubic yards of sand that flows down the river each year could not make its way past the jetty and to the shores of Camp Ellis.[133] The beach could not be replenished, and the edges of the neighborhood began to erode. In all, Camp Ellis was losing several feet of land per year. The sand that broke away in the current was pushed north to the towns of Old Orchard Beach and Scarborough. Meanwhile, to the south, a neighborhood called Biddeford grew from the sediment coming down the river.

Troubled, the town told the Corps to come fix the problem they had created. The Corps widened, lengthened, shortened, and otherwise modified the jetty in an attempt to balance the navigational needs of the river and stop the damage in Camp Ellis. These efforts continued for a *century*, but the erosion continued.

The decisions individuals and institutions make manufacture disaster[134] and across the country our risk is growing. When I was growing up in Maine, I had no idea that just down the road there was a microcosm of the trends increasing our global risk, but Camp Ellis is just that.

When Reverend Ellis first built his house, it was surrounded by acres of grassland. It was a wooden house with no plumbing or electricity, which could only be reached by a single dirt road. Despite sitting back from the shore, the possibility of flooding was ever present, as you might expect when a house is built at the intersection of a river and an ocean. But there was bountiful natural protection to minimize the severity of the flood risk. When those protections were overwhelmed, rebuilding did not take long because there wasn't much to be damaged and only basic infrastructure to be rebuilt. This, combined with the strategic location of the land near the winter port, justified building there despite the risk. Today, that natural protection is all but gone and dozens of modern homes and hard, complex infrastructure has been built, which has made the flooding more severe, more damaging, and more expensive.

The story of Saco reflects a familiar, global reality: naturally vulnerable places are often located where people need or want to live. And today, there are simply more people living in more dangerous places. In Saco, residents congregated around the two most vulnerable parts of town: the ocean and the river. In spite of the erosion, Camp Ellis continued to grow as development along the shore was driven by the fishing and tourism industries. Visitors were drawn to the seaside neighborhood, conveniently located a few miles down shore of Old Orchard Beach, a popular summer spot hosting the likes of Louis Armstrong, Duke Ellington, Charles Lindberg, and New England's elite.[135] For locals, Camp Ellis proved to be a reprieve from the industrial downtown area, which saw its population double in the mid-1800s.

The center of town continued to develop along the river as it provided the necessary power and transportation for the mills. In March of 1936, downtown Saco was mercilessly flooded as ice jammed on the river. The Saco River overflowed its banks, filling businesses and homes across town with water. Residents recovered and continued to grow the town to today's population of about twenty thousand people.

Not many make a living off the sea anymore, but the pull of the water continues, as new houses are built regularly. Now, the river is primarily used for recreation and the mills have been converted into lofts and art studios, but as recently as 2019 there were proposals to build new luxury apartments—aptly named The Waters—and businesses right at the river's edge despite the obvious flood risk.[136] There are similar issues along the rest of the Maine coastline as the population density is anticipated to continue increasing in the coming decades.

Over 40 percent of US residents live on the coast, a number that has nearly doubled since the 1970s.[137] We have built towns and cities at the foot of volcanoes for the fertile soil, next to the ocean for fishing, and by rivers for transportation. Now more people than ever live in these places. Compared to highly urbanized places, particularly megacities, the scale of Saco's risk is tiny. The complexity and magnitude of risk faced by cities like Jakarta, Delhi, and New York City is staggering.[138]

Flooding is to be expected when you live beside the ocean. For some, ocean views and the ability to walk to the beach outweigh that risk. Others stay because it is where their families have always lived, or because the water is tied to their livelihood and way of life. As someone who grew up along the shore, I get it. You can show me all the flood-risk maps in the world but the coast is home. It sounds ridiculous but I felt claustrophobic living in landlocked states.

During one visit home to Maine, I searched the books housed in the town library's basement for old photos of Camp Ellis. They show fields of dunes running the length of the peninsula with only a few scattered houses protruding from the landscape. In the past when storms made it up the coast, the dunes acted as natural protection along the shores, absorbing much of the storm surge and minimizing or even entirely eliminating the damages. But when the jetty was built, the dunes began to erode. The old dirt roads that used to absorb water were paved over. The natural protection disappeared and Camp Ellis's vulnerability to flooding increased.

The development of these places where people want, or need, to live has itself increased our risk. Unchecked development and industrialization has led to detrimental impacts on the environment, including the destruction of ecosystems like the dunes in Camp Ellis that once served as natural mitigation. Barrier islands that protected the shores from storm surges were once free to shift and grow as currents changed, but once people built on them, they had to be held in place. Entire mountains were deforested, destroying the root systems that literally held the land together and prevented mudslides when it rained. Shores that were lined with sand dunes, wetlands, and swamps, which took the brunt of storms and minimized coastal erosion have been paved over, leaving no way for water to be absorbed into the ground. So at the same time more people moved to these increasingly vulnerable places, the natural ecosystems that once provided protection were destroyed.

In the past, when Camp Ellis did flood, residents could quickly rebuild their homes, but today's recoveries are more complicated.[139] The costs and technical skill required to rebuild modern homes can elongate the recovery period. Washed-out roads cannot simply be replaced with new dirt. Instead, repairing damage to roadways, wastewater, sewage, and electrical systems can take months and hundreds of thousands of dollars, if not more. This complexity contributes to the rising costs of disasters.

The tools of industrialization themselves are a source of risk. Nuclear power plants dot our landscape, pipelines run through our drinking water, drilling platforms litter the horizon, dams loom above towns, trains full of explosive oil run through the middle of cities, and factories spew chemicals into the air. Each on their own can lead to disaster, but when they interact with natural hazards, like storms, the potential for destruction is amplified.

This collision between nature and technology is at the heart of some of this century's most devastating disasters. In New Orleans, it was the failure of the poorly constructed and maintained flood infrastructure that caused the most devastation; if the levees had not been breached, hundreds of lives would have been saved.[140] The presence of complex infrastructure and technology means more can go wrong, in more ways, and with greater consequence.

These factors are why disaster sociologist Dr. Dennis Mileti said disasters are "by design." The hazard—that is, the actual hurricane or earthquake—may itself be a natural phenomenon, but the location of a neighborhood along the shoreline or developers who have not built homes to withstand an earthquake represent decisions made by individuals, organizations, and governments. How and where we build, what our laws and regulations stipulate, the resources we have, and other decisions that we make before, during, and after the event manufactures our risk. Disasters happen when we fail to manage risk, and when decisions are made that increase our vulnerability.

Some have understood this for a long time. After the Lisbon earthquake of 1755, which killed seventy thousand people, philosopher Jean-Jacques Rousseau introduced a social science interpretation of the disaster.[141] In a response to French writer Voltaire, Rousseau argued, "nature did not construct twenty thousand houses of six to seven stories there, and that if the inhabitants of this great city had been more equally spread out and more lightly lodged, the damage would have been much less and perhaps of no account."[142]

Dr. Gilbert White, the father of floodplain management, famously wrote, "Floods are 'acts of God' but flood losses are largely acts of man."[143] Disaster researchers have tried to explain this for decades.[144] In recent years, I have begun to hear survivors wrestle with this distinction in the aftermath of disaster. They explain how the flood risk was known, but the town could not afford to build a levee, or how regulations were waived that opened the door for development in high-risk floodplains. Survivors correctly attribute fault to decisions made in, or for, the community.

This hopefully signals the beginning of an important shift in the public's understanding of disasters. Correctly naming the cause of disasters is a necessary first step in being able to hold accountable those at fault. Further, it lays the groundwork for us to be able to effectively minimize, or even eliminate our risk. This reality, that human decisions are at the center of creating our risk, is actually good news. If we are the cause, we can also be the solution.

In Saco, residents understand the repetitive flooding in Camp Ellis is not natural, but rather manufactured through decisions made in the past and decisions being made now. It also means that there are still things that can be done to prevent continuous, or permanent, disaster in Camp Ellis, which is why residents there are fighting for their future.

The Fight for Camp Ellis

AS THE YEARS have turned to decades, Camp Ellis has continued to wash away. The neighborhood has lost nearly three football fields' worth of land since Reverend Ellis first built his house. In the fifty years since the Corps last worked on the jetty,[145] eight streets including thirty-five houses have been lost to the waves.[146] It's Maine's very own Atlantis.

When friends come to visit me in Maine, I take them to Camp Ellis between trips to the lighthouses and the Old Port. We walk out onto the jetty where I show them the causes and consequences of over a century's worth of bad policy decisions. We walk through the neighborhood, and I point out the hodge-podge of efforts residents have tried to keep the water out. There are some obvious adaptations, like the houses closest to the water that tower on stilts twelve feet high. Less visible are the homes elevated only a few feet on cinderblocks. Front yards are lined with garden walls and beach grass—not aesthetic choices but rather an attempt to keep homes dry. Once you know to look for them, you can find adaptations to the flooding everywhere.

Windows have storm shutters, plywood leans along the side of houses ready to be hung at a moment's notice. At the end of each street corner the town leaves out the barricades they use to shut the roads down when the flooding begins. While residents have done what they can to protect their homes, they have also turned to local government for help as the situation worsens.

Millions of local dollars have been spent on response and recovery efforts in Camp Ellis. Over the years the town has tried to minimize the damage to the neighborhood. They've built the thirteen-hundred-foot riprap seawall, just a giant pile of rocks to match the jetty. The wall creates a barrier between the edge of the neighborhood and the rugged waves. Beneath the beach the town installed a Geotube, essentially a gigantic sandbag.[147] They have also tried to maintain the remaining natural protection, making beach nourishment and dune construction projects a part of regular maintenance. Each of these projects has bought Camp Ellis some time and lessened some of the damage caused during storms, but their combined efforts have not stopped the erosion or flooding altogether.

Walking through the neighborhood, I point out the other adjustments that could be made. More houses could be raised and others moved farther back on their properties to provide a few extra feet of safety. A higher seawall could be built. A different type of pavement, one that absorbs water, could be installed, along with other types of drainage systems. These wouldn't be permanent solutions but they could minimize the damage a bit more and for a bit longer. The problem though is bigger than these projects. It's bigger than what homeowners can fix on their own and it's more than what the town can manage. The town of Saco has already done what they can afford on their own. Local resources and patience have dwindled so despite the increasing damage, local options have run out.

Back when the town first realized the problem with the jetty, they demanded the Corps of Engineers take responsibility for the erosion in Camp Ellis. Surprisingly, the Corps admitted that building the jetty contributed to the erosion in Camp Ellis and that they were responsible for helping find a solution. Over several decades, the Corps produced dozens of studies and proposals for what to do about the jetty but did not implement any of their proposed solutions.[148] Instead, it felt like the Corps had decided on a strategy of keeping the town busy with a never-ending stream of studies, proposals, paperwork, and meetings because waiting them out was cheaper than actually solving the problem.

If this was their intentioned strategy, it's worked. Saco did not have the time to wait for the Corps to find a solution, because Camp Ellis kept flooding. As they've negotiated with the Corps, Camp Ellis residents and the town government undertook the variety of small-scale mitigation efforts like the Geotube in the hope of holding the flooding off until the Corps comes through with a more permanent solution.

As this process has continued on for decades it feels like the Corps of Engineers is gaslighting the town. It does not seem to matter that they admitted fault because the town has no leverage to hold them accountable. Still, residents cling to the Corps' admission as the evidence that the federal government should pay to save Camp Ellis. Federal financial support is also, likely, their only option.

Recently, one of my trips home to Maine happened to align with a town meeting on the Corps' latest proposal: the addition of a seven-hundred-fifty-foot spur onto the jetty. I joined the residents of Saco at town hall, an old brick building with granite stairs and a clock tower that looms over Main Street. A hundred townspeople, tanned and in flip-flops, milled among friends and familiar faces while they waited for the meeting to begin. To my right, a woman propped up an iPad to take notes.

The woman to my left sat down with a three-ring binder over-filled with paperwork—a recovery binder.

The mayor called the meeting to order. His stark white hair and bright green shirt stood out in the somber crowd. The tension in the room was familiar to me. It mirrored the energy of town hall meetings I had been to across the country, filled with those who had found themselves in a similar position to Camp Ellis residents. I have found, though, cordiality doesn't last long when the government stands in the way of people's homes and futures.

The mayor announced the others in attendance. The city administrator, representatives from three-fourths of Maine's congressional delegation, Saco's representatives, the harbor master, a representative from the Corps, and other volunteers and officials each stood as their names were called. The mayor concluded his opening remarks by saying that they were all here with their best wishes and thoughts. His words got right to the tension of a decades' long fight, where priorities and resources clashed. It was a word of warning disguised as a pleasantry.

Tension aside, the town had at least gotten everyone in the same room. Industry, media, representatives of every level of government, and residents had, if not a seat at the table, at least a seat in the room. This was a good sign, because often when mitigation projects are not implemented, it is because they lack community participation.

Like a game of chess, each player had come with a strategy, anticipating the moves of the others in attendance. I had chosen a seat in the very back of the room so I could see it all unfold. Some residents at the meeting had dedicated decades of their lives to this fight. This was not a low-stakes community meeting about traffic complaints or petty neighborhood arguments; this meeting could determine the future of Camp Ellis.

The first speaker noted the high attendance at the meeting was likely driven by the storms a few months earlier. This renewed interest could be used as momentum to find a solution. As he alluded to the "window of opportunity" that the latest storms had caused, my guard went up reflexively.

While the language of "windows of opportunity" has often been co-opted by disaster capitalists in the wake of disaster, there are instances where local communities can make the changes that they need and want post-disaster. Political scientist Dr. Thomas Birkland has studied how some disasters lead to the implementation of mitigation or other policy changes. These disasters, known as "focusing events," must capture the attention of both the public and policymakers, be sudden and dramatic, relatively uncommon, and affect a concentrated geographic area.[149]

The most recent storms in Camp Ellis certainly seemed to have captured the attention of the public and policymakers, given the high attendance at the town hall meeting, so I hoped this would be an opportunity for change. At the same time, I was skeptical meaningful action would be taken. A century of storms had not brought change, and I had no reason to think this latest storm would be any different. Still, as I sat there, I let myself wonder if this would finally be the meeting that secured Camp Ellis's future. It is intoxicating, after all, to think you will get to witness history.

A representative from the Corps spoke next. He gave a long, technical, and unnecessary explanation of an upcoming river dredge. The Corps would suck sand up from the riverbed and deposit it along the shores of Camp Ellis. This would replenish the beaches that had been gutted by the jetty. I recognized this immediately as a distraction tactic. They were replenishing the beach because they already had the sand from the dredge on hand—a long-planned project unrelated to the recent flooding. Further, replenishing the shore was at best a short-term solution. The insinuation though was that the project was being done because of the erosion in Camp Ellis. The crowd also seemed to see through this tactic and grew increasingly impatient.

The other shoe dropped quickly. The sand would not last long, according to the Corps. They admitted that in some places the sand would go as fast as they pumped it while in some areas it could last four to five years depending on storms, which, to me, sounded like a generous estimate. As the rest of the room caught up someone asked incredulously, "So it's a Band-Aid?" Finally, in a moment of honesty, the representative from the Corps replied, "Completely."

Attention finally turned to the real reason everyone had come—the proposal to add a spur to the jetty. The spur would be another rock structure that extended seven hundred and fifty feet in length from the jetty, and would hopefully reduce the erosion along the beach. Town officials began by clarifying the spur would not eliminate flooding entirely, but emphasized that it was the best available option. Earlier that day, Saco had submitted their final report to the Corps in which they agreed to the spur project. This wasn't a done deal though. Having the town sign on was only an initial step to actually having it built. The Corps would still need to give their final approval for the project and then sign a formal partnership agreement with the city. There seemed to be confusion over the actual costs of the project. On the high end the number $71 million was thrown out. This would cover the cost of the spur and decades' worth of beach renourishment. Upon hearing the price tag, one woman in the crowd wondered aloud if they could just push some boulders down to the jetty themselves.

Back in 2007, Congress authorized $26.9 million, unattached to a specific project, to mitigate the erosion in Camp Ellis[150] and Senator Susan Collins was able to add another $5 million. In the years since, Saco has spent $2.5 million of that money on efforts related to Camp Ellis, but the rest of the money has just sat waiting for the Corps and the city of Saco to agree on its use.

The spur project would need to be incorporated into the Corps' overall 2019 budget, which would then be sent to

Congress. Residents would need to lobby their congressional delegation—Senator Susan Collins (R), Senator Angus King (I), and Representative Chellie Pingree (D) to make sure the project was not removed from the Corps' budget. Assuming the Corps' proposed budget remained intact, it would be added to the overall national budget that Congress would need to pass and President Trump sign. If Congress instead passed a continuing resolution, the project would not be funded. The audience scoffed. In the summer of 2018, it seemed unlikely Congress would agree on anything—including a budget.

The hits didn't stop there. Assuming Congress approved the budget and the president signed off, there was still the matter of actually building the spur. The earliest construction could begin would be two years later, in the spring of 2020. Given the two storms from that year alone, this timeline wasn't comforting. Erosion and flooding wouldn't stop while Congress negotiated the budget and construction began. If Camp Ellis didn't flood in the interim it would be due to nothing but luck.

There was one final slap across the face. The Corps would agree to the spur project, but in exchange the town would have to agree to relieve the Corps of all future responsibility. The cost of any and all future maintenance related to the jetty would fall to state and local government. Absolving the Corps was an unpopular concession locally, for obvious reasons, and agreeing to the spur project had major implications for Saco's local budget. The town was already spending an annual average of $300,000[151] on flood-related expenses in Camp Ellis, but that number would increase if Saco agreed to the spur. So, although the overall cost of flooding in Camp Ellis could decrease with the addition of the spur, the proportion of the cost being paid for by the town would likely increase.

It was a bad deal, but it was also the only deal being offered. The Corps had the town backed into a corner. Saco was being forced to sell their future for a quick, partial fix. City officials understood the problems with this plan and that they didn't know how they would afford the future expenses but, they argued, it was still the best option given the prospects of future flooding, and the absence of a good alternative.

Frustration and desperation rippled like a wave across the audience. It didn't take a doctorate in emergency management to see that it would be difficult, if not impossible, to clear all of these hurdles, especially because there wasn't much residents could do. The process was largely out of their hands. It was up to Congress to agree to the funding or to pass the budget. At most, residents could call and plead with their Congressional delegation to make sure it made it into the final budget. Someone in the crowd noted that the power of bringing baked goods to Senator Collins's office should not be underestimated.

As the attendees began to grasp the full extent of the situation, an exasperated man in the audience leaped up and yelled what was on everyone's mind: "How long is too long? Another thirty-three years?" He reminded the room he had been fighting with the Corps to fix the jetty for three decades. He paused to shout "NOW IS THE TIME" throughout his impromptu, impassioned speech infused with the confidence of someone who had done this many times before. He was met with resounding applause.

As the agitation in the room became palpable, the limits of government support were tested. The possibility of suing the Corps of Engineers, an idea that had been discussed in the past, was raised again by someone in the audience. The town could pay $25,000 for Saco to obtain an attorney in Washington, DC, and begin legal action. After all, the Corps had admitted fault. City officials, including the mayor, brushed the idea aside, stating they would have to raise taxes to fund what would undoubtedly be a lengthy legal battle.

The threat of a lawsuit also prompted the representatives of the Congressional delegation to get to their feet for the first time. They warned that if the city pursued a lawsuit against the Corps, the delegation would no longer be allowed to talk with the town about the jetty and would stop advocating for Saco in Congress. Mainers' favorite new chant, "shame on her," rose up from the crowd directed to Susan Collins's staff. It seemed like it would take more than some homemade whoopie pies for Senator Collins to support the town's efforts.

Ultimately, the idea did not move forward, as others noted the rarity of municipalities winning lawsuits against the federal government, particularly against the Corps of Engineers. Someone near me mumbled something about New Orleans. They may have been right that the Corps was responsible, but that didn't mean they would win—and certainly not in time to save Camp Ellis. A legal battle would drag on for years and residents knew they did not have the money, or the time.

The meeting came to an abrupt end, with no real resolution, and without much confidence in the future of the spur. With only half of the needed funding for the project, and the looming political barriers to receiving more funding, the possibility of this final jetty modification was daunting, if not impossible. There was a time, decades ago, when Camp Ellis had many options for how to manage the flooding, but as years passed, the options dwindled. Saco has negotiated with the Corps of Engineers for decades, and in that time there has been no substantive action taken to prevent the erosion and flooding in the long term. It seemed Saco would continue to be on its own.

A year later they began dredging the Saco River. As promised, the Corps piled new sand up along the shore of Camp Ellis. That summer my brother and I went for a walk through Camp Ellis to see the sand added from the newly completed dredging project. We stood on the jetty and looked out over a rebuilt beach standing between the neighborhood and ocean. It looked good.

On Halloween, just months later, I went for another walk around Camp Ellis. It was raining and leaves were falling, but it was an uncharacteristic sixty degrees. The rocks lay bare against the waves; most of the sand was gone. The Corp's expected three-to-five-year solution had lasted closer to three to five months.

There was one giant elephant in the room during the town hall meeting that no one brought up: the inevitable future of the Maine coast. If climate change had been mentioned, the already minimal effectiveness of the Corps' spur project would have been called into further question. In the seventy years Saco has argued with the Corps over the jetty, another, more dire problem arrived, one that makes the situation much more daunting—climate change.

Disaster Denialism

CAMP ELLIS OVERLOOKS the Gulf of Maine, which begins off the coast of Cape Cod and covers the area along New England and up the Canadian coast. It is the water Rachel Carson wrote about and I grew up swimming in. The Gulf of Maine is the fastest warming body of water of its size anywhere in the world.[152] The lobster population is changing,[153] the cod are dying,[154] the whale population is declining,[155] and the sea is rising.

When scientists talk about climate change they usually use "global averages" to describe the changes. This can be slightly misleading because it gives the impression that all places will be affected uniformly. While it is a practical approach to communicating a complex, global problem, it is also deceiving because it can obscure the actual risk to specific communities. In reality, some places will experience more sea level rise more quickly than the global averages indicate. On the coast of Maine, climate change is not a future concern, its effects are already being felt, and Camp Ellis has been hit with a climate change trifecta:

sea level rise, more intense storms, and an increasing threat of spring river flooding.

In the best-case scenario, Camp Ellis will see a two-foot rise by 2100, but in what many scientists consider the worst-case scenario, the neighborhood will all but disappear completely under a six-foot rise.[156] At eighty years from now, 2100 is a lifetime away, but as I learned on the coast of Louisiana, climate change is not a switch that flips on January 1, 2100—the seas are already rising.

As the oceans rise, the warmer water can invite stronger storms farther north.[157] The more frequent, stronger storms are amplified by sea level rise bringing higher tides and storm surges into the streets of Camp Elis. At the same time, increased precipitation and changing temperatures alter the timing of snow and ice melting, which can threaten river flooding. Exactly how much land and how many homes will be lost in Camp Ellis, due in part to these changes, depends largely on global climate action. The damage caused by two feet of sea level rise versus six is vast—it is the difference between Camp Ellis being permanently above water or not.

Even if we were not in the throes of a climate crisis, and Camp Ellis enjoyed the climate Reverend Ellis did when he first built his home, the neighborhood would still face significant flooding. Their land would still be eroding because of the jetty and some nor'easters would still strike the coast, but the long-term prognosis would not include a possible six feet of sea level rise and such frequent storms.

The climate once known to Reverend Ellis doesn't exist anymore. In Camp Ellis, and across the country, we have spun ourselves into a web of risk. Camp Ellis demonstrates how it is not climate change alone that creates our risk, but instead weakens what is already fragile. The same systems that are vulnerable to climate change, such as food and water, are simultaneously undermined by our failure to maintain infrastructure, reduce population growth in hazardous places, and regulate industry. The same factors that have increased our vulnerability to disaster are what have led to the climate crisis, which has, in turn, further increased our risk. Climate change isn't just a future threat anymore. The crisis has begun, and its first mission is to amplify the risks we have long faced.

Along the coast from Eastport, Maine, down around the Florida Keys and across the Gulf of Mexico to Texas and back up to the tip of Alaska, people are already facing these exact same issues and worse. Camp Ellis is just one little neighborhood, in one little city, in one little state, but their struggle is similar to what communities across the country are beginning to face.

One challenge is working around the people who deny our risk. Most obvious are the climate deniers—those who do not believe or understand that the climate is changing because of human action. Some are passive in their denial but others, funded by the Koch brothers and others from the oil and gas industry, actively work to prevent climate action.[158]

Then there are the people who know the climate is changing but engage in behavior that ignores its inevitable impacts. The coastal real estate market is booming as millionaires are buying condos in Miami. The city of Houston is allowing new neighborhoods to be built in high-risk floodplains. Developments on the West Coast are pushing into the Wildland Urban Interface and into the path of wildfires.

Then there are the deniers who believe the climate is changing but do not believe human action is the cause. The day following Hurricane Irma's landfall in Florida, Tom Bossert, spokesman for the Department of Homeland Security, said, "We continue to take climate change seriously, not the cause of it, but the things we observe."[159] A version of this type of denier also misidentifies the causes of disaster.

In the midst of the 2018 Camp Fire, the deadliest in California's history, then President Trump took to Twitter to place the blame on "poor forest management" and threatened to withhold disaster aid.[160] Experts including disaster researchers, firefighters, government officials, and journalists were quick to correct him.[161] His claim obscured the multiple factors driving increasing wildfire in California—climate change, development, and the US approach to fire suppression.[162] Raking the forests, as he suggested, isn't a solution to these problems.

As advances in science have shaped our understanding of what causes disasters, we have become better at predicting our risk. We generally have a good idea of the likely threats to our communities. This local knowledge is formalized in risk assessments made by emergency managers. These assessments account for the hazards a community may face and consider which characteristics of their communities may make them more vulnerable to a given hazard. It is these risk assessments that form the foundation, at least in theory, for what a community can do to minimize or prevent disaster, what we in emergency management call mitigation. Successful mitigation requires us to correctly identify the source of our risk. In the past, we did not have the science to help us consistently and correctly understand the root causes of disasters, which made mitigating that risk difficult or even impossible. Now, we know a lot more. Correctly identifying the sources of our risk is exactly why acknowledging and understanding climate change is fundamental to effective emergency management.

It is also why misplacing, obscuring, and denying the cause of disaster is so dangerous. In attributing blame to nature or the supernatural, we avoid interrogating our role in creating these risks. When then President Bush claimed we could not anticipate the levees breaching in New Orleans,[163] or when Trump misattributed blame for the California wildfires, we moved further from understanding and managing the conditions that led to the occurrence of these disasters. In the aftermath, we fail to address the actual source of our risk, which in turn, undermines our ability to prevent future disasters. Like climate denialism, we must also contend with this disaster denialism.

Fortunately, for all those who deny disaster in its many forms, there are others, like the residents of Camp Ellis, who are well aware of their risk, understand the long history of decisions that were made to create it, and are trying, desperately, to do something about it.

Technically, the flooding in Camp Ellis isn't a disaster. Not yet. The persistent flooding has thus far been an "emergency." Emergencies occur when damage is confined to a relatively isolated geographic area and the immediate needs of those affected can be met relatively quickly.[164] Help doesn't converge on the community in the same way it does during disasters and catastrophes. Instead, local people and organizations use their experience and training to take action in a planned, coordinated way. Making this distinction is not meant to minimize the experience for the people who live there but it does provide context for understanding why the situation has unfolded in the way it has.

I can easily imagine a future flood on the coast of Maine that becomes a disaster. It would come as a hurricane wandering farther north than usual in the warming waters of the Atlantic. High tide would bring coastal flooding worse than the Patriot's Day storm. Towns running the length of Southern Maine, including Saco, would be inundated. Camp Ellis would all but drown beneath the storm surge. With the entire coastline flooded, the town would be overwhelmed. Neighboring towns would not be much help as they focused on their own recovery. The state, with the rest of coastal New England, would call on the federal government to help and it would take years to rebuild. As the climate continues changing but the jetty remains unchanged, the possibility of this disaster becomes more and more likely.

Even if this storm does not come, Camp Ellis still has a problem. This is what climate change forces us to contend with—not only catastrophes but also persistent instability. Emergencies can build over time. The regular flooding and ongoing erosion in Camp Ellis drains our attention and resources even as it doesn't quite tip the scales to become a disaster. The turmoil instead becomes a daily reality.

A Reactive Approach

DURING THANKSGIVING BREAK in 2018, I did my usual drive down Beach Street to Camp Ellis. There was a gentle snow falling, but the wind whipped hard enough for me to favor the shelter of my car instead of walking through the streets. I saw no one else as I made my way slowly through the neighborhood. Winter had arrived, signaled by the absence of out-of-state license plates, boarded-up windows, and extra concrete barriers positioned timidly between delicate homes and an agitated sea. It had been eight months since the last storm and North Ave was rebuilt, front yards unburied, decks replaced, and houses repainted. The cleaned-up neighborhood was a testament to a determined town, unwilling to lose any more ground.

Upon closer inspection, though, the facade of the recovered quaint seaside neighborhood fell away to reveal Camp Ellis's fragility. They are as vulnerable as they are resilient. My fears for their future returned as I noticed the newly repaved roads dipped in places, creating pools of standing water, a small sign that signaled a probable drainage problem. One of the homes, raised twelve feet in the air, had a new for sale sign out front. I stopped here and there to roll down my window and take photos for my students.

I was in the middle of teaching a course about mitigation at North Dakota State University, and had been using Camp Ellis as a semester-long case study of a place that is trying to prevent future disasters. In class, I had been teaching my students not only about Camp Ellis but also all of the ways we can protect any community from any hazard. Most disasters are preventable, or their severity can at least be significantly minimized. Learning how to do this is a fundamental part of emergency management.

We can build flood walls and levees, landslide walls, lava flow channels, and do storm water management. We can do risk mapping and require public risk disclosures. We can deny services to people who build in high-risk areas, and change land-use policies, building codes, and zoning. We can do wetland restoration, reforestation, controlled burns, and sand replenishment to restore our natural protection. We can raise buildings above flood lines, strap down hot water heaters, and secure bookcases to walls. We can change affordable housing policies and enforce environmental regulation. The list of things we can do to prevent disaster is nearly endless.[165]

We have always had to adapt to our environments, and because of that, we actually know a lot about how to mitigate our risk. Recently, there has been growing interest in mitigation, as more people have begun to experience the effects of climate change. Some communities have long understood their futures are at risk but many more are just starting to understand that their futures hinge on the implementation of mitigation measures (often called "adaptation" in climate circles).[166] Recognizing the need for mitigation may seem obvious, but it's an important first step.

All of our experience, and research, tells us that we should want to do mitigation—to take a proactive approach to emergency management. We know mitigation can save lives, minimize physical injuries, and lessen disaster-related mental illness. It also can save us a lot of money. A study from the National Institute of Building Science found that for every $1 the federal government spends on mitigation, $6 is saved in response and recovery.[167]

Yet despite understanding the benefits of being proactive we have, historically, been reactive in how we manage disasters. We wait for disasters to happen and then deal with them, rather than doing everything we can to prevent them from happening in the first place. If you're thinking to yourself, "hmm, that seems like a bad approach," you would be correct.

As Thanksgiving break came to an end I decided to take one last drive through Camp Ellis. Being a disasterologist can turn you into a bit of a pessimist, and my students were known to leave my class with a bit of an eye roll and protests of "this is so depressing." So I was excited to bring my students an update on the recovery progress from the January and March storms. My students are the ones who will be responsible for implementing mitigation measures in the communities they will eventually work in, and they needed to see some progress, even if it was just some newly paved streets. This small piece of good news would be a good note to end the semester on! The joke was on me, though, because just twenty-four hours later Camp Ellis was underwater again. So much for optimism.

My photos were outdated by the time I landed in Fargo the next day. What local meteorologists referred to as a "standard storm" had formed along the coast. It was so un-noteworthy that coastal flood advisories had not even been issued. Yet the storm, coinciding with high tide, pushed the waves up over the beach, and easily brushed aside concrete barriers. The tide surged into the neighborhood and covered the fresh pavement with a new heap of sand. North Ave, which the town had just spent

$250,000 to rebuild, was in pieces again. In just a few hours, the neighborhood was right back to where it had been just eight months earlier. There would be no Presidential Disaster Declaration for this one. Not enough towns in York County experienced damage, so they would not be eligible for federal funding. Saco was once again completely on its own.

The new storm certainly had not made the news in Fargo, so I began class with a slideshow of the recovery photos I had taken. I felt a twinge of guilt as I lulled my students into a false sense of believing things in Camp Ellis were going well. After months of learning about this little neighborhood on the other side of the country some of them had become invested in its future, so they were eager to see the progress.

In class they had been learning about what research tells us about which communities are able to successfully implement mitigation measures.[168] One reason I had selected Camp Ellis as a case study for our class was because the neighborhood aligned almost perfectly with what research tells us is required for a community to prevent disaster.

Saco residents have been engaged in the fight for mitigation from the start. They've taught themselves the process and understand the barriers they face. They've formed committees and held town hall meetings. The man who had yelled "how long is too long" at the town hall meeting had, decades earlier, founded a local organization, Save Our Shores, with the mission of organizing residents to advocate for Camp Ellis. In 2019, the organization was launched in its new iteration SOS Saco Bay, with the expanded mission to organize residents' responses to the erosion in Camp Ellis and Southern Maine more broadly—a perfect example of coalition building.

The flooding and these efforts have received regular media attention in glossy New England magazine features, local news coverage, and op-eds in the *Portland Press Herald*. Despite being a small state, Maine's representatives have advocated for Camp

Ellis for decades and local representatives and elected officials have been engaged in the process. Despite Camp Ellis aligning almost perfectly with what research tells us a community needs to be able to successfully mitigate their risk, the likelihood of a permanent solution for Camp Ellis wanes by the day.

I cleared away the recovery photos and pulled up a livestream filmed the day before by a local meteorologist who had been in Camp Ellis, which showed the streets filled with water. My students watched as the decades of erosion, a storm, and higher than usual tides made their way into the neighborhood. They knew, as did I, that it was inevitable that Camp Ellis would flood again, but that didn't make it any easier to watch.

The problem is that there has not been enough mitigation fast enough to make a difference. Camp Ellis residents have done what they can as individuals to minimize their risk, and they have taken action through local government to implement community-wide measures, like the Geotube and riprap wall. These efforts have been impressive, but they are simply not enough to keep the neighborhood above water in the long term.

Residents in Camp Ellis have done everything they were supposed to do, and yet they were still underwater. The work they've done, spanning decades, is noteworthy. The dedication, resources, and expenditure of energy is impressive. Many communities do not, or cannot, do what Camp Ellis has done.

Camp Ellis is the poster child for a community that has spent decades working on the same problem with no permanent solution. The process of identifying a community's risks, sorting through the technical options for how to mitigate those risks, organizing local leadership, getting community buy-in, and securing funding, through actual project completion does not happen quickly or easily. Despite how far they still have to go, residents of Camp Ellis had already accomplished so much. The fact is, though, that they are up against the clock.

Efficiency and timeliness are fundamental to the successful

completion of mitigation projects. Disasters do not stop while a community figures out how to navigate mitigation programs, and every minute a project is not implemented is one where the community remains vulnerable. We know timing matters, but unfortunately that runs in direct opposition to the process of applying for federal funding, which is complex and time-consuming. The longer it takes to implement mitigation, the greater the likelihood that more damage is done. Further, the longer a community goes without the needed changes, the more difficult it can become to find a technical solution. Conditions evolve so the original mitigation plan may become obsolete. Having a plan to prevent a disaster doesn't do you any good if your house floods again before the plan is implemented. Places like Camp Ellis are out of time.

Of all the places I have seen confronting climate change, Camp Ellis scares me the most, because they haven't been successful, even with the system working in their favor. It may not feel like it when you stand in the aftermath of the latest flood, but Camp Ellis is faring far better than many other places around the country. They have a plan in hand for a project that could minimize their future risk, and the opportunity to move forward and seek support from the federal government. This alone required decades of political negotiation and technical study. They have an engaged community, with local leadership that understands their risk and wants to find a solution. They are supported by local government, have bipartisan federal representation fighting for them, and the Corps of Engineers is on record as admitting fault, rather than "just" blaming Mother Nature. They are organized, well educated, and have the time and personal resources to dedicate to this decades-long battle. Plus, their demographic profile is favored by the system: a white, middle-class neighborhood in a financially stable town and state. Camp Ellis has had a century-long head start in a system built for communities exactly like theirs, and yet they are barely above water.

In class, we wrestled with the reality of how much better off they were than many communities and what the experience in Camp Ellis can tell us about how government will, or will not, support mitigation efforts in marginalized communities—especially as our national risk grows in the face of the climate crisis. It is painful to watch places like Camp Ellis become trapped in these cycles of destruction, especially when the blueprint for avoiding disaster exists. Yet across the country, we are slow to implement these proactive changes that we know will keep people safe and minimize future costs.

Tens of millions of dollars is not an easy sell for a project that will buy a single neighborhood a few more years. And here lies the problem. It doesn't matter if you've found a technical solution or how much community, media, and political support you have if you can't get the check to cover the costs.

As of this writing, there is no resolution to the problems in Camp Ellis. Town officials consider there to be three options left for the neighborhood. First, the town can stay the course and continue negotiating with the Corps of Engineers over project details (the current situation). Second, the town can pursue legal action against the Corps, an option that would likely be expensive, lengthy, and ultimately unsuccessful. Or third, they can abandon Camp Ellis.

The Future of Camp Ellis

I HAVE NEVER met anyone who wants to leave Camp Ellis, but they could. The town could give up on the jetty project and the other temporary fixes. They could decide that future projections are too grim for a viable, sustainable solution. They could decide they want—or need—to leave. The government could buy their homes.[169] The town could implement land-use policies that would prevent future construction where the neighborhood once stood. They could turn the neighborhood into green space, an extension of the neighboring public beach or nearby state park. This would permanently eliminate the flooding problems in Camp Ellis and have the added benefit of extra protection for the homes that sit farther inland—an important consideration, given that sea level rise threatens to bring the shore closer to these other properties.

People make the difficult decision to leave their homes all the time. People who have lost everything in a wildfire decide it would be easier to move and start over somewhere else. Some have the decision made for them when they cannot afford to rebuild after a tornado. Sometimes entire neighborhoods, or even entire towns, decide their risk of future disaster is too high and relocate somewhere safer together. People move when all other reasonable options are exhausted—when there is no other choice.

During the 2020 presidential election, for the first time, climate change was a focus. Seventy-eight percent of Democrats called it a top issue, and a number of climate town halls were held during the primary.[170] The questions candidates fielded largely centered around emissions and green jobs, but every now and then they'd be asked about adaptation—"what about the people who are being affected by sea level rise?" The candidates all had similar answers—"managed retreat."

At some point the risk of living in certain places, like along the coast, becomes unsustainable as the climate changes and our risk grows. People will have to move away from the coast, and many argue it would be better for that exodus inland to be managed in an effort to minimize pain and suffering. (At the risk of stating the obvious, the impacts of climate change are also felt inland, so moving away from the coastline is not the be-all and end-all solution it is often treated as.)

Politicians often talk of managed retreat as though it is a new strategy that will need to be used only in rare circumstances in the far-off future. In reality, it is a strategy we have a long history of using and are already actively doing. What they call "managed retreat" is what we in emergency management call "buy-outs" and the federal government has been funding them for decades. Since 1989 FEMA alone has funded forty thousand home buy-outs across the country.[171] When the cost of living somewhere becomes too high, local governments may advocate for buy-out programs. Homeowners receive a payout for their home and the government prevents anyone from building on that land in the future. While simple in theory, this approach quickly runs into a number of problems.

First, homeowners have to agree to buy-outs. Most of us like where we live, and we do not want to move. Some of us have lived in the same town our whole lives. It's where our families are from, and where our friends are. We have a connection to the land. Our jobs are there. Even when we know objectively that the most cost-efficient solution is to move, we can often lull ourselves into believing that the next disaster will not come.

At the town hall meeting in Saco, one woman shared that five generations of her family grew up on the now disappearing beach. Another shared that he had spent his entire adult life working to fix the jetty, but did not think he would live long enough for the project to be completed. This was intergenerational pain. It is easy to sit at a think tank in Washington, DC, writing policy recommendations that houses along the coast should not be rebuilt because the seas are rising, but it's much harder to stand in front of the people who live in those houses at a town hall meeting and tell them that they have to abandon home.

Assuming homeowners do agree to move, the logistics quickly become difficult. All the same problems—the need for media coverage, political support, technical plans, and money—that come up with mitigation projects like adding the spur to the jetty have to be dealt with. Navigating agreements and funding mechanisms can take years. In 2019, the National Resource Defense Council (NRDC) released a report that found buy-out programs take an average of five years to complete.[172] Further, because the buy-out option often comes about post-disaster, homeowners go through this process during recovery. In fact, it leaves some in a state of recovery limbo for an undetermined amount of time.

In the years between their home being destroyed and knowing they will be bought out, something that is not at all a guarantee, homeowners struggle to make recovery decisions. Should they spend their money on rebuilding their flooded home, even though it could be torn down in a few years? Or should they not rebuild and cross their fingers that a buy-out eventually comes through? If they choose not to rebuild, where will they live until they actually get the buy-out money? If they decided not to rebuild and then the buy-out falls through, then what do they do? Start the rebuilding process years after the disaster? Abandon the home?

Even once the government signals they will agree to a buy-out and it looks like there will be funding, more questions must be addressed. How much money will people receive? Exactly which houses will be included? When will people need to move? Where will they move to? When will the properties be destroyed? What will be done to ensure nothing is built back in their place?

There are also discrepancies in who has received buy-out funding. An NPR investigation found that FEMA disproportionately funds buy-out programs in white communities as compared to communities of color.[173] Again, this seems to be the result of bias built into the design of the program's application process. Money is allocated based on a cost-benefit analysis that prioritizes more expensive properties rather than those who have the greatest need. Further, wealthier communities may have more resources to lobby for mitigation measures and better connections to decision makers. A similar issue has been found related to some types of wildfire mitigation as well.[174]

This is particularly frustrating because often it was housing, transportation, and development policies that pushed people of color and marginalized groups into these more physically vulnerable areas in the first place. Then, the risk in these neighborhoods is further compounded as they receive less investment in mitigation measures (e.g., drainage, infrastructure maintenance). When a disaster does occur, they are likely to have fewer resources to respond and are forced into the discriminatory recovery process, as we saw in New Orleans.

When a town looks at which parts of the community should

be turned into green space, these neighborhoods are the most obvious candidates because of the frequency of flooding, lower home values, and residents who are less likely to have the resources and political power to fight back—exactly like the Green Dot Map. These federal programs don't just fall short of meeting the needs of communities across the country, but they can perpetuate discrimination and actually deepen inequality. This process highlights the very valid fear that some people and some communities have about their future. It raises the question of justice. Which communities will receive help and which will be sacrificed to save others?

The buy-out option hasn't even been seriously discussed, at least publicly, in Saco. The federal government could afford to buy out the homeowners in Camp Ellis, but as we look to the future and see the number of homes in the path of flooding, there are serious questions to be asked about the sustainability of this approach. By 2040 climate change is expected to contribute to repetitive flooding in fifteen hundred communities along the US coast, including major cities like Miami and New York City. Are there enough resources for all these affected communities? Who decides which homes are bought out and when, and how much will they get? Already in places like Houston, the list of people who want to be bought out is longer than the resources available.[175] In the past thirty years FEMA has moved only forty thousand households. Can we effectively, and justly, scale this up? And soon? This is why when politicians are asked what they will do about sea level rise, shrugging and saying people will "have to move" isn't a sufficient or realistic answer.

For most people, and most communities, moving is the last resort. Yet we have to face the reality that it is the direction we are headed in as the climate crisis worsens and other mitigation measures are not implemented in the interim. The clock is winding down and our poorly woven safety net isn't going to catch everyone who needs it unless we act quickly.

There's one house in Camp Ellis that stands out from the rest. It's not one that is precariously close to falling into the ocean or one towering twelve feet in the air. The house that always catches my eye is a little cottage a few lots inland, which gives it the appearance of being more protected than most. A little wooden sign is affixed to the sea-worn front window box in the summer and moved inside the front window during the winter months. In swirling, hand-painted letters it reads Aunt Alice's Cabin 1903-1990. Moving would not just mean admitting defeat against the Corps and nature, it would mean abandoning a neighborhood that holds entire lifetimes, generations of memories, a whole culture, and summers at Aunt Alice's house.

aps. As I worked out which towns were under each color on
e map, the tug in my chest tightened.

Climate writer Mary Annaïse Heglar has described the times
he can envision the disasters that climate change will eventually lead to as "climate vision."[177] I experience a version of
this—perhaps "disaster vision"—in the moments just before a
disaster. It's a reflex that came free with my doctoral degree in
emergency management.

Harvey played out in my head days before it actually happened. I could see the destruction Texas was barreling toward.
I could see the houses that would fill with water and the cots
in shelters that would fill with people. I put placeholders in for
the faces of the people who would die and tried not to think
about the pain it would bring the people they would leave behind. For a moment, it's like being haunted by ghosts who are
not yet dead.

Sometimes these visions flash by in just minutes, but sometimes they linger on for days. This is always the hardest part for
me because there is nothing in the moment that I can do to stop
it from happening. I spiral through thoughts of all that could
have been done to prevent the disaster from happening if only
elected officials had acted earlier. It can also be lonely waiting
for everyone else to understand what is about to unfold.

Harvey made landfall on the Texas coast on August 25 as a
Category 4 hurricane.[178] Five feet of rain fell across East Texas
before the storm continued farther east for a final landfall in
Louisiana. So much rain fell that the National Weather Service
had to add two new colors—dark purple and light purple—
to the maps in order to accurately reflect the rainfall totals.[179]
Harvey broke the US record for the most precipitation from a
hurricane.[180]

The storm was exactly as devastating as I expected it to be
from the moment I first saw the multicolored maps. Harvey, the
largest hurricane to make landfall in the US since 2005, required

PART THREE

Response: The Disaster

"Cow."

—HELEN HUNT, *TWISTER*

I WAS DEEP in the Maine woods at my family's cabin on Rangeley Lake at the end of August 2017. I was there to get some writing done in peace and spend the last few days of summer on the lake. In the evenings I drove into town to use the Wi-Fi from the local coffee shop. On one of these trips I sat outside on a bench, watching the haze of the late afternoon sun settle over the lake. As my laptop slowly connected to the Wi-Fi the online world came rushing in.

Meteorologists were in a frenzy. I had to pause to catch my breath as I saw why. Twitter was flooded with maps of Texas covered in dramatic waves of deep orange, red, purple, and blue. Hurricane Harvey was brewing off the Yucatán Peninsula, but faced an unobstructed path over a warmer than usual Gulf of Mexico.[176] It would collide into the Texas coast and Houston. I felt a tug in my chest. This one was going to be really bad. I pushed aside the work I had intended to do that week and moved into the town coffee shop to monitor the evolving hurricane the most significant response since Superstorm Sandy in 2012. The blank faces I had imagined days earlier were filled in as the photos of the seventy people killed in the storm were published in local news outlets.[181] Thirteen million people were affected and over forty thousand took refuge at shelters.[182] Hundreds of thousands of buildings were damaged[183] and an estimated million cars were destroyed.[184] Damages topped $125 billion.[185]

I would have watched this storm regardless of where in the world it was headed, but I had just finished writing my doctoral dissertation on flooding in this exact part of Texas so I was particularly attuned to the damage to come. Harvey would challenge any state, but I knew this was more than Texas could handle. In the space between what Texans would need and what local emergency management could provide were thousands, if not millions, of people who would suffer. Texans would be quickly overwhelmed and would need help from neighboring states and the federal government.

East Texas had four presidentially declared floods in the two years preceding Harvey.[186] The Memorial Day Flood of 2015 brought eleven inches of rain. The remnants of Hurricane Patricia in October 2015 brought nine inches. The Tax Day Flood in April 2016 brought seventeen inches and two months later, a flood in June 2016 brought eighteen inches. These numbers paled in comparison to Harvey's sixty inches of rain, more than three times as much as these previous floods.

It is most often the big disasters, like Harvey, that capture our attention, because their death tolls are high and their economic impacts vast. But I think it is the comparatively smaller, often overlooked disasters like the Tax Day Flood, that are most instructive about our future. The response to these persistent, repetitive, disasters that are bigger than the flooding in Camp Ellis but smaller than Katrina, wear communities down and, in many ways, are more challenging. They come with relatively few deaths and little national media attention. They do not get special bills in Congress to help survivors rebuild. The donations dry up quickly and volunteers don't come to help a decade later. They are the disasters that largely leave local communities on their own to survive.

A Quick Response

ON A DECEMBER morning in 1917 a man named Samuel Prince, curate at St. Paul's Church, was at breakfast when an explosion rocked Halifax, Nova Scotia.[187] Prince had helped search for and bury victims from the Titanic just a few years earlier, and once again he took action to help the victims of disaster. An accident involving a ship carrying explosives had caught fire and exploded in the harbor, sending a shockwave of destruction through the seaside city. The impact killed two thousand, injured another nine thousand, and initiated the field of disaster research that we have today.

Prince left Halifax to study sociology at Columbia University.[188] There, he wrote his dissertation on the response and recovery to the explosion. Prior to the publication of his work, writings on disasters had simply given an accounting of what had occurred.[189] Prince's dissertation took a different approach. The study outlined a series of observations about human and organizational behavior and has come to be regarded as the first systematic social analysis of a disaster.

In the decades after Prince's work, there were a handful of social science studies done on disasters, but it wasn't until the late 1940s, in the shadow of WWII and the outset of the Cold War, that the field exploded. The federal government was interested in how Americans would react to a nuclear attack on US soil, so they funded researchers to study human behavior during times of crisis.[190] Sociologists traveled from one disaster to the next—chemical fumes in Donora, Pennsylvania; plane crashes in New Jersey; tornadoes in Arkansas; earthquakes in Alaska; and floods all across the country—using them as a proxy for an enemy attack.[191] In doing so, throughout the second half of the twentieth century, disaster scholars like Dr. Enrico Quarantelli, Dr. Russell Dynes, Dr. J. Eugene Haas, and Dr. William Anderson laid the groundwork for today's body of research on how people react to disasters.

Among their main findings is that in times of crisis communities come together.[192] Contrary to popular belief, we do not panic, loot, or descend into chaos when disaster strikes. Rather than widespread antisocial behavior, we see again and again that people respond prosocially.[193] We help each other. People, groups, and organizations that did not plan on becoming involved in a disaster, and who often do not have any kind of emergency management training or experience, find ways to help. Survivors and volunteers who have only just met coordinate their efforts, as they work together to address the most pressing needs even before help arrives from the outside.[194]

And outside help does arrive. People and aid converge from surrounding communities as outsiders learn of the disaster. The amount of assistance that arrives can be staggering, with researchers going so far as to call it a "mass assault" of help.[195] It is impossible to get an accurate count, but some estimates are shocking. Ten percent of all Americans reported donating or volunteering post-9/11[196] and at least forty thousand volunteers spontaneously descended on Ground Zero in New York.[197] Researchers found that 60 percent of San Francisco residents and 70 percent of Santa Cruz residents volunteered in some capacity following the 1989 Loma Prieta earthquake.[198] This happens internationally as well. An estimated two million people volunteered after the Mexico City earthquake of 1985[199] and as many as 1.4 million people volunteered after the 1995 Kobe earthquake in Japan.[200]

These unplanned, often untrained and informal helpers join with trained responders and government resources to save lives and manage the disaster.

Much of the data necessary to study disasters can be gathered long before or after a disaster, but some data is perishable. During the response, disasterologists have only a small window of time, sometimes just a few hours, when we are able to interview survivors and responders and make important firsthand observations before the information is lost forever. The challenge, as it has always been, is to get to the site of a disaster in time to do this work.

We can easily identify probable disaster scenarios in advance—Miami will be hit by a hurricane, there will be an earthquake in California, and a tornado in Oklahoma—but we do not know when exactly each of these will happen. One tactic is to hedge our bets by designing a study ahead of time and then waiting for a disaster to occur that meets the parameters. This was the situation I found myself in during the spring of 2016 as I worked on my doctoral dissertation.[201]

Although Prince, Quarantelli, and the other founding disaster researchers established that help flows into disaster-affected communities, in the decades since surprisingly little research had been done on who those people are, what they do, and why.[202] Figuring this out was the purpose of my research during the Tax Day Flood. Knowing more about the volunteers who help during disasters can help us better facilitate their efforts. This, in theory, could maybe one day make our response to disasters, as a whole, more effective. In fact, because volunteers are such a fundamental part of how we respond to disasters, our lack of understanding about them and the work they do is a troubling weakness in our knowledge of emergency management.

Since I was studying how volunteers react during the *response* to a disaster, I needed to be there to witness the disaster myself. What emergency managers call response is what you see under the "breaking news" graphics on CNN and in Hollywood disaster movies. It is when life-saving measures are taken to protect lives, property, and the environment.[203] Response begins when we learn a hazard is about to happen. If there is time, evacuation or shelter-in-place orders are issued. Shelters open and hospitals fill up as search and rescue efforts get underway. The actual response to a disaster usually does not last long. Life-saving tasks can end in as little as a few hours, and rarely last more than a week, but, because it's dramatic, response is what tends to capture a disproportionate amount of media coverage and our attention.[204]

In mid-April, I saw my chance. I took one look at the expected rainfall totals for Houston and texted my dissertation adviser that I would be on the next flight from North Dakota to Texas. I raced home, packed my bag, booked a flight, and left for the airport.

I had not been the only one watching the weather report for Houston. When I landed in Dallas on a layover, I found teams of volunteers in matching T-shirts representing the major national disaster nonprofits waiting at the gate. Like I had, they understood the serious flooding that was coming and were trying to get on the ground as soon as possible to help with the response. I was happy to see them because volunteers were exactly who I was going to Texas to study.

There are many ways to categorize the people who volunteer during disasters, but generally we make a distinction between affiliated volunteers, meaning those working with a specific organization, and spontaneous volunteers—those who just show up to help on their own.[205] Affiliated volunteers often work with well-established disaster nonprofits. As is the case in recovery, nonprofits fill the gap between what individuals can do on their own and what government does to respond to an unfolding disaster.

The Red Cross is the most well-known, but they are only one of the groups that respond when disasters strike. The Salvation Army, Islamic Relief USA, Team Rubicon, Catholic Relief Services, among other organizations, regularly respond to disasters across the country. Their volunteers tend to be well trained and have prior disaster experience. Many of these groups even coordinate their responses with local emergency management officials. In any given disaster some combination of these groups come to help, and the Tax Day Flood was no exception.

These affiliated volunteers aren't hard to locate for interviews because their association with an established organization makes them easy to spot. It's the spontaneous volunteers, who work on their own, that take a bit of detective work to find. There is often no accounting of who volunteered, what they did, or even an accurate count of how many of them were there. It was because of these spontaneous volunteers that I would need to be there during the disaster. If I didn't go around the community and find them myself, it would be impossible to locate them later.

I had a head start in getting to Texas, but, as evidenced by ongoing flight delays, the weather was already turning. Getting to the site of a disaster can be difficult. The speed and ease with which outside help, and disaster researchers, can arrive is complicated by destroyed or otherwise impassible roads, bridges, ports, and airfields. So as the first evacuations and search and rescue began in Houston, I was stuck at the Dallas airport eating a Caesar salad and casually browsing the airport bookstore.

Just as discussions about renting cars to drive the rest of the way became serious, we got lucky and the weather cleared long enough for us to land in Houston. As we came through the clouds, passengers leaned over one another to see out the windows. Beneath us, muddy water escaped the bounds of rivers and bayous. Houston is so big that even in the worst floods, you can stand in one part of the city and not know a disaster is unfolding just a few streets away. From above though, I could see the water weave itself from one town to the next, revealing the extent of damage. I looked around the plane and wondered which passengers on our flight would be returning to flooded homes.

A Warning

IN 1900, GALVESTON, a barrier island running parallel to the Texas coast, was the third busiest port in the country and with a population of more than forty-thousand had been dubbed the New York City of the Gulf.[206] In early September, the US weather bureau in Washington reported a storm in Cuba. At first, bureau employees stationed in Cuba said it was nothing Texas should worry about (despite disagreement from their Cuban counterparts). Soon though, Isaac Cline, Galveston's meteorologist, received a telegram about a tropical storm in the Gulf and began looking for signs of an approaching hurricane. At the time knowledge about these storms and technology for tracking them was still in its infancy. Cline himself didn't believe Galveston could be affected by a hurricane and had previously argued against building a seawall to protect the city. As the hurricane approached the island conditions quickly changed. Just a few hours before landfall Cline went against protocol and issued a hurricane warning, but there was little time for action.

The hurricane arrived in Galveston where the wind and waves

ripped wooden houses apart, nail by nail, until they were re-fashioned into wooden mountains. If warnings had been clearly given early and heeded, residents could have evacuated. Though the property damage would have been the same, it would not have left as many as twelve thousand dead, earning the storm the title of America's deadliest hurricane.[207]

It has long been an important goal of researchers to increase our warning times. Some hazards come with no notice, but others come with minutes, hours, or even days' worth of warning. In theory, the more warning the better because it provides precious time for people to take action to protect themselves.[208] Beginning in the mid-twentieth century, advances in our understanding of hazards coincided with advances in technology and opened the door for more effective warning systems. Earthquake early warning systems can give people just a few seconds to step away from an unsecured bookshelf.[209] A tornado siren can give people a few minutes to get into their basement. A few days' lead on a hurricane can be enough time to evacuate an entire city. This has all come with a dramatic decrease in the long-term trend of natural hazard-driven deaths globally, which is especially impressive considering the population growth experienced in the past seventy years.[210]

After the levees failed in New Orleans, the city vowed to prevent another death toll like Katrina's from ever happening again. This vow came with changes to the city's evacuation plans so that no one who wanted to leave would be left behind.[211] In August 2008, Gustav became the first serious storm to threaten New Orleans since 2005 and was the first real test of the city's changes. It also happened to be the same week I moved to New Orleans.

As Hurricane Gustav made its way across the Gulf, I found myself living in an unfamiliar city with absolutely no plan for how to handle a hurricane. At the time, I was eighteen, naive, and largely unaware of the danger I was in, but my parents, a thousand miles away, certainly understood. They called screaming and told me in no uncertain terms that I was to get on a plane back to Maine immediately. Naturally, I decided against it.

I am embarrassed to admit I had moved to New Orleans without ever considering the possibility of being affected by a hurricane myself (yeah, I know). I had done exactly nothing to prepare for a hurricane. I did not know the first thing about reading a hurricane map or even what categories meant. I hadn't been in New Orleans long enough to learn which streets always flood or how far inland I would need to go to stay dry. Baton Rouge? Alabama? Arkansas? Even if I had known where to go, I did not have a car and flights were filling up fast. Basically everyone I knew in Louisiana had also just moved to the city days earlier and were in a similar situation. Then there was my checking account. The sum total of my high school summer job had left me with just enough money for textbooks and cheap wine—not a week-long evacuation, which, I quickly realized, is surprisingly expensive.

The idea of evacuating at first felt dramatic, unnecessary, and too complicated to figure out. So I decided I would stay put and weather the storm on campus. I lived in a cinder block dorm room on the tenth floor of a building with hurricane shutters on the windows. If there is a building that can withstand a hurricane, Buddig Hall at Loyola University is it.

In my defense, sometimes the safest option is to stay put.[212] In situations like tornadoes and chemical releases, sheltering in place can be safer than trying to evacuate. A quick glance at Gustav's cone of uncertainty, however, suggested this was not one of those times. I began to second guess my plan as I watched the storm grow stronger and closer. If we lost power, the tenth floor in hundred-degree temperatures did not seem habitable. Even if we stayed dry in our building, how long would we be stuck there? What would we eat? How would we call for help if we needed it? And who would we call?

These questions became moot when the New Orleans mayor ordered a mandatory evacuation of the city, stating honestly, "We cannot afford to screw up again."[213] By the time the university announced they were shutting down and we all had to leave, flights were booked and I had limited options. Fortunately, one of my new friends who lived on my hall was from Des Allemands, a small town in southeast Louisiana, and knew exactly how to handle a hurricane evacuation. She took pity on me and three other non-Louisianans in our dorm. If she was annoyed with our ignorance, she hid it well.

The five of us piled ourselves and our most treasured belongings into her little Mazda3 and turned our evacuation into an adventure. We were among the last to leave campus, joining two million others in the first mass exodus from New Orleans since Katrina. Even with the contra-flowed interstate, the traffic turned the usual five-and-a-half-hour trip to Alabama into a twelve-hour trip. When we finally arrived in Opelika the five of us packed into a cramped hotel room that would be our home until the hurricane passed. We watched from there as Gustav barreled toward Louisiana.

FEMA estimated only ten thousand people remained in New Orleans during Gustav, far fewer than the number who stayed during Katrina.[214] Since Katrina, the city had created better procedures for providing public transportation for those in need. The city had become more diligent about issuing evacuation orders in advance. These numbers did seem to indicate that at least something had been learned from the previous evacuation failures, even if there was still room for improvement.[215] In 2009, for example, the organization Evacuteer formed to help the city with evacuations. They have put statues up throughout the city to designate evacuation pickup locations and undertook a public education campaign.[216]

In the end, Gustav turned and left New Orleans largely un-scathed. After a few days we were able to return to the city and the entire situation became fodder to tell our friends back home. In the absence of another catastrophe befalling the city the entire experience turned into just a wild college story. Now, though, I see it as a lesson in vulnerability.

Warning the public is hard. Researchers in disciplines like psychology, sociology, and communications have found a long list of factors that affect our decision-making when faced with an impending hazard.[217] Officials themselves need to receive forewarning of the risk and be able to send a warning to the public, which itself is a complicated endeavor. For the warning to actually be worth anything the public needs to actually re-ceive it. They need to be able to understand the warning and agree that the threat exists. This means they have to trust the source who sent it. They may also look for environmental cues or confirm the information they've received with others around them before taking the warning seriously. The warning itself needs to be accurate and the public must receive it in time to be able to act. Of course, they also have to know how to act. What actions do they need to take to protect themselves? They also have to have the skills, resources, and ability to complete those actions. This often all has to happen very fast, and condi-tions may change quickly. Everything has to line up just right for people to effectively protect themselves.[218] At any point in this process response can be derailed, and lives put in danger.

Not everyone has the ability to avoid disaster like I did. Al-though I hadn't done anything to intentionally prepare for a hurricane, I was in the end more prepared than I thought. I had a support network to lean on and parents who could send me money for this unexpected expense. I was able to make up for my lack of local knowledge by finding someone who knew what to do. I was able to leverage this to find transportation and a place to stay. I didn't have anyone I needed to care for or a job that I could lose while I was away. We also got lucky that the hurricane missed the city and we were able to return without the need to do any kind of recovery.

There are plenty of stereotypes about the people who don't evacuate during a hurricane.[219] There's the guy who stays out of defiance, ready to take the hurricane on with gun and flag in hand. Then there are the people who stay because they've "been through worse before" and some who want to stay to protect their property from the looters they swear will show up. If these people are making an informed decision and they understand help will not come to save them, then that is their choice.

I'm much more worried though about the people who want to leave and can't.[220] Even if someone receives a warning and understands there is a danger, they may not be able to act. They may not have anywhere to go or a way to leave. They may have a physical disability, or some other legitimate reason that prevents them from taking the recommended actions to protect themselves.[221] This is who emergency management has the greatest responsibility to help.

Sometimes, from the outside looking in, it can be easy to judge the decisions being made by the people who are experiencing—or are about to experience—a disaster. But often, people are making decisions without having all the information or resources they would need to make a better choice. Generally, people make the best decisions they can with the information and resources they have available to them.

Mass evacuations are grueling. Moving millions of people in a matter of hours requires extensive planning, supportive infrastructure, clear and constant communication, adequate resources, and a little bit of luck. Fortunately though, most disasters do not require millions of people to move across state lines. There was no mass evacuation during the Tax Day Flood in Houston, but many people did have to leave their homes.

On my way to Texas I scrolled through social media posts, sent emails to contacts based in Houston, and monitored official government communications. I watched as street after street went underwater in real time. I had a map of the city printed out and marked down the areas where I saw flooding reported. Yet, as I watched the disaster start to unfold online, I had a hard time figuring out how many people in Houston saw the flood warnings and were able to get to safety. The warnings were staggered and changed constantly over the course of many days as the flooding worsened. Monitoring the warnings required constant vigilance.

In Houston many found evacuation wasn't an easy option because the roads and their cars were already flooded by the time they realized it was necessary to leave. Once they were able to make their way out of the flooded areas some didn't have anywhere to go. In most disasters, relatively few people stay at shelters. Most stay with friends or family or they are able to rent a hotel room.[222] In Houston, however, thousands of people needed a place to stay, so shelters were opened around the city. Many people arrived via city buses, wet from having waded through the rain and flood water.

There was one picture I saw circulating online that showed a group of people floating an air mattress through the floodwaters. Sitting on the mattress was a woman holding on to two small children.[223] I thought of a line in a poem by Warsan Shire. The image felt so familiar to me. It could have been taken during any number of floods in the past fifty years, but this time I couldn't look away.

This particular flood did not happen quickly, unfolding instead over many days. The city had so much experience with flooding that knowing which areas would flood shouldn't have been a surprise. I think this is why I struggled with the photo of the kids on the mattress so much. The family should have been warned in plenty of time to leave before the streets filled with water. The photo signaled a disconnect between warnings that gave me enough time to fly across the country, but not enough time for people to evacuate their homes to shelters just a few blocks away before the streets filled with water. This picture, to me, meant that something had gone wrong.

A Planned Response

I HAD ALWAYS thought of New Orleans and Miami as the flood capitals of the South, but it turns out Houston rivals them both. East Texas has long experienced flooding, but in recent years the floods have been more damaging and more frequent.[224] Even before Harvey, Harris County, which encompasses Houston, had incurred more property loss, more flood-related deaths, and more urban flooding per capita in the past forty years than any other US county.[225] Even the flooding is bigger in Texas.

Houston has written the playbook on how to flood a city. The flood risk in Houston has been created in the same ways it has been in most places. In recent decades, the population of Harris County has increased by millions, coinciding with a development boom. As a result, Houston now has significantly more people and more structures in an area that was already vulnerable to flooding. Entire neighborhoods have been built in places that are known and expected to flood. Local officials have cleared the way for development, with few regulations or building codes that account for flood risk. In the process, much of the natural environment that at one time absorbed rainfall has been destroyed. Just as Houston grew into a sprawling concrete metropolis, the climate changed and has brought increasing torrential rainfall to the region. Plus, their hurricane risk is changing as storms become more intense, and as sea level rise helps storm surges push farther inland.[226]

Some local officials have gone as far as to deny their risk. According to the *Texas Tribune*, as of 2016 the Harris County Flood Control District refused to study the impact of climate change on the increased rainfall, and insisted the city's policies encouraging development were not increasing flood risk.[227] New buildings were not built to withstand flooding and existing buildings were not updated to account for their new reality. Some people did not even know they were living in, or buying homes in, areas with a high risk of flooding, because the flood maps were not correctly updated. The city also has not done enough to maintain the current flood infrastructure system—which is made up of a series of natural bayous, dams, drainage channels, levees, sewers, and the like.[228] The system cannot keep up with the rate of water making its way through the system during these high rainfall events. When the rain comes, there's nowhere left for the water to go, except into Texans' living rooms.

The risk is growing, and without acknowledgment of why the risk is growing, successful mitigation can't happen. All it takes is the right hazard, at the right time—or, really, the wrong time—and disaster strikes. As I say, same story, different disaster. It was against this background that the Tax Day Flood, like the Memorial Day Flood before it and Hurricane Harvey after it, took place.

When a disaster happens, government officials get on TV to hold press conferences. If you look closely you'll see that some of the people there are wearing polo shirts and khaki pants, usually the kinds with lots of pockets. This is the uniform of the people who work in emergency management agencies, and most people don't know much about them. These emergency management officials stand by as mayors give rallying speeches about the city coming together to address the disaster. When they finally get the microphone, they rattle off those strings of numbers—how many are evacuated, how many are staying in shelters, how many have been rescued—to describe the disaster.

During the Tax Day Flood, news of which areas had flooded and the actual scope of the damage constantly evolved, but the long and short of it was that many neighborhoods in Houston and surrounding communities were underwater.[229] Over eleven hundred homes were flooded across multiple counties. One neighborhood in Houston, Greenspoint, where some New Orleanians had relocated to post-Katrina, was particularly affected.[230] Eighteen hundred high-water rescues were conducted and around a dozen shelters were opened. In the end eight people were killed and there was $5 billion in damage.[231]

Press conferences held in the midst of disaster can at times devolve into acronym soup, as emergency managers rattle off the various agencies that are helping in an effort to show they're coordinating the response and that things are under control. When the press conferences end, emergency management officials go back to their emergency operation centers (EOCs) where all the other emergency management staff who you never see on TV are hard at work. Some EOCs have walls lined with monitors and look like something out of a disaster movie, but others are hidden away in old basements of town halls and aren't more sophisticated than a large conference room.

For decades, government has worked to improve our approach to managing disasters.[232] We have frameworks, plans, systems, structures, and procedures that outline roles and responsibilities. There are trainings, drills, exercises, education and degree programs, and certifications to teach emergency managers everything they need to know. We have an entire discipline of study[233] and an emerging profession.[234] We have hundreds of emergency management agencies all across the country and organizations whose entire missions are dedicated to managing disasters. These efforts have certainly improved our responses to disasters but in many ways there is still much work to be done.

In 1900, Clara Barton, at seventy-nine years old, traveled from Washington, DC, to Galveston to oversee the Red Cross operations firsthand.[235] Barton had founded the American Red Cross (ARC) in 1881 after spending time in Europe learning about the International Committee of the Red Cross. She saw a need for the organization in the US and it grew quickly. The Red Cross was first chartered by Congress in 1900, which put the organization in the unique position of being an independent nonprofit but also having the lead responsibility for response operations in the United States.[236]

Before emergency management agencies even existed, the Red Cross was the primary national driver of the response to disasters, like the one in Galveston. Although their role in disasters has greatly diminished in more recent decades as they are no longer the lead response organization in the US, they are still expected to help. Long before a disaster actually happens, Red Cross officials often work with emergency management officials to establish what the organization will be responsible for during disaster. They train and prepare for responding to a variety of scenarios.

Groups like the Red Cross expect to be involved when disasters strike, because responding to disasters is part of their organizational mission. They know they have responsibilities in the overall response and understand how their role integrates into the larger response system. They expect to see each other at the scene and have an idea of what each of the other groups is supposed to be doing. Among them, they have a wealth of disaster expertise and experience. They have the resources they expect to need and know where and how to get more.[237] They are often connected and tied in with the government agencies. Among the disaster organizations that respond the Red Cross is arguably still the most prominent.

There is some variation from place to place, but generally the Red Cross is responsible for opening shelters and providing food and water to survivors during response. Which is why I made the Red Cross my first stop in Houston. I arrived at their headquarters downtown, and a volunteer working the front desk led me upstairs to their emergency operation center. There was an eerie stillness as we walked through the building. We wound our way through the halls past empty dimly lit rooms that stood in contrast to the crisis outside. We walked upstairs and a door opened to bright lights and the organized chaos of the command center.

The volunteer didn't know where to find the person I was supposed to meet so she left me to search on my own. I wandered through the room, a large office space with fluorescent lights and folding tables. Signs hung from the ceiling distinguishing the different tasks the volunteers sitting below them were working on. I picked one side of the room and started interviewing volunteers about how they had ended up there helping and the work they were doing. I also listened in on various briefings to try and gauge where in the city I would find volunteers working—especially spontaneous volunteers.

From these briefings I learned the updates on official numbers and what local government was doing. Government-led relief efforts were primarily focused on high-water rescues, evacuations, and blocking off flooded roadways to prevent people from driving through standing water. As I went from meeting to meeting within the Red Cross operation center, I found the assessments of what was happening outside varied. This kind of conflicting information made the work I was trying to do more complicated, but it wasn't unusual. During a response there is a lot of information flying around very fast. It comes in from different sources and changes quickly, not to mention the game of telephone that's played as it goes from one person to the next, all while communication networks may not be fully operational. This means getting a grasp of even basic numbers—like how many people are at shelters—can be difficult, if not impossible.

Despite the conflicting information, from Red Cross headquarters, it looked like the quintessential disaster response. Various agencies and organizations appeared to be coordinating with one another. Backup, in the form of the national disaster nonprofits I had met at the airport, had arrived. Everything that should have been happening seemed to be happening.

Despite the many policies and agencies that have been created to support disaster response, this system is imperfect. Although efforts have been undertaken to implement standardization, it can look and operate differently from place to place. Further, it can be easily derailed and overwhelmed—the system itself is not always very resilient.

The actual response is more complicated than what is communicated in dry, safe emergency operations centers. The appearance of a totally controlled and coordinated response can dissipate once you get outside. As I sat in the Red Cross headquarters hearing about how everything was going well, I couldn't get the image of the kids on the mattress out of my head. I noticed a list of shelters written on the wall and wrote down the addresses. The kinds of volunteers I needed to find weren't hanging out in emergency operations centers—they would be out working among the survivors.

An Unplanned Response

ARMED WITH THE official list of shelters from Red Cross headquarters, I headed to the largest one. Inside, I found a long line of people waiting to check in. They carried an assortment of duffel bags, trash bags, and other odd containers filled with whatever they had grabbed before leaving their homes. A police officer hovered a few feet away, looking on as the line grew. Volunteers assigned cots for the night, explained various rules about coming and going, and noted when meals would be served. Exhaustion was the prevailing feeling; other emotions would come later.

I slipped back outside. No one noticed or questioned my presence. It was too late in the evening for interviews and as the adrenaline wore off I realized I was also exhausted from traveling. I would return the next day to start talking with the volunteers.

The next morning, I arrived at the shelter to find much of the same. I had been given permission to be there, but I wanted to introduce myself to the "shelter lead" and explain I would be making observations and interviewing volunteers. On my arrival, I approached the volunteers sitting at the entrance who had replaced the women from the night before, and asked them to point me to the shelter lead. I did not have a name, just a position title. The two volunteers consulted each other. They decided the person walking around with the clipboard was probably in charge. They did not know the person's name and had not seen them since the previous afternoon. They did rattle off a list of other volunteers by name, and were able to tell me exactly where I could find them. I noticed they made a point of emphasizing none of them were with the Red Cross.

This particular shelter had been a point of extensive discussion at the Red Cross headquarters the previous day. Those discussions, however, did not align with what I found once I was there myself. There were a few specialized medical volunteers, but otherwise the person with the clipboard, whose name no one seemed to know, was the only person from the Red Cross. I encountered the other dozens and dozens of volunteers had no Red Cross affiliation or training, and were locals who had spontaneously arrived to help. They were running the kitchen, checking people in, handing out supplies, cleaning, and playing with the kids staying at the shelter. It was days before other trained Red Cross reinforcements arrived from as far away as the Pacific Northwest to help run the shelter.

I spent the day watching and listening to the volunteers: their experiences, their frustrations, their concerns, their anger, their excitement, and the satisfaction they felt for helping their neighbors in a time of need. These sentiments were repeated as I made my way around East Texas in the following days. I heard from volunteers about the work they were doing and how they ended up helping. Throughout the week, I talked to some volunteers who were affiliated with various disaster nonprofits, but the majority I met were not working with a national group. They were just locals who showed up to help their neighbors.[238]

★ ★ ★

At another one of the shelters I visited, I interviewed a volunteer who was running the kitchen. On paper, the shelter was run by the Red Cross, but he wasn't a Red Cross volunteer. He told me word had gotten out around town that this particular building would be a shelter. People began showing up from the nearby flooded neighborhoods looking for help before any officials or volunteers had arrived. Unsure of what to do, a building manager had unlocked the doors and invited everyone inside. Then he started calling around asking people he knew to come help. The guy who was now running the kitchen had been one of those phone calls.

When he arrived to help, he found the kitchen empty. He had never run a kitchen before, nor had he ever worked in a shelter, but he saw hundreds of hungry people so he figured it out. For the next five days, he kept hundreds of evacuees and volunteers fed. Impressed, I asked him how he pulled off such a feat. He laughed and said he was just treating it as though it was a wedding. He had some experience working for an event planner and had figured out he could apply those skills to running the shelter. There were some bumps in the road, like a visit from the health department with a list of health code violations. He took it all in stride though, improvising as he went.

I heard similar stories from volunteers across the city. Most had heard about the severity of the flooding from their friends and family, on social media, or from news reports. They arrived en masse at the shelters and other hubs of relief activity, like donation distribution sites. They would spend some time wandering around to see how things worked and then they jumped in to help. They found something that needed to be done and they figured out how to do it. Some had volunteered during previous floods or had been through flooding before themselves. Most volunteers had no formal disaster training, but it didn't seem to matter much. Anyone can make a sandwich.

The volunteer who ran the kitchen mostly credited the steady stream of volunteers and donations for their success. Employees from local restaurants came to the shelter, often unannounced, with meals ready to go. People from the community had dropped food off, too. He organized volunteers to sort donations and to help prepare and serve meals. When there was a lull in donations he called around to local businesses and asked for more help. Fortunately, many local businesses had been unaffected by the flooding and were able to help.

As I wandered throughout East Texas, I found a lot of volunteers spending their time managing donations. In one shelter, I saw a few people getting off an elevator. I hadn't noticed at first, but there was a second floor. I got in and rode up with two little boys in the midst of an imaginary cops and robbers game. I was hit with the distinct smell of a thrift store as the doors opened. The boys took off running and I looked up to see a giant atrium filled with dozens of folding tables piled high with mountains of clothes. A few women dug through one of the piles while the two boys from the elevator chased each other in circles. I scanned the room for volunteers and found only one.

She told me the year before during the Memorial Day Flood, her house had flooded. She stayed at a shelter like the one we were in now. When she saw on the news that shelters were opening again, she dropped everything to come help. She knew what it was like to have to stay in a shelter and how important volunteers were in running them. As we talked about how she had ended up coming to this particular shelter, I noticed she never stopped folding and sorting the clothes on the table. I sensed that being back in a shelter and seeing the city flood again was hard for her, and folding the clothes was a welcome distraction.

When she had arrived that morning, she asked around looking for something useful to do until another volunteer told her there were clothes upstairs that needed to be sorted. She went up and developed a system of sorting the piles by size and gender. She did not know where the donations had come from, but she thought they had just been dropped off at the shelter. At the same time that people arrive in an affected community to help, there is also a convergence of donations, a trend that has a long history.[239]

In 1900, when news broke across the country of the hurricane in Galveston, help started making its way to Texas. One peculiar donation arrived—boxes of high heels for just the left foot. What luck for anyone in Galveston only missing one shoe! One hundred years later, these useless donations continue. An aid worker reported that in the days following the 2010 Haiti earthquake, they received a donation of a single rollerblade.[240] After the Sandy Hook shooting people donated stuffed animals to the children of Newtown, Connecticut, by the tens of thousands. There were enough toys to fill a sixty-thousand-square-foot storage space.[241] These are extreme examples, but they reflect the fact that many donations, though often well intended, fail to match with the needs of survivors.

It's not only that these donations may not be helpful, but that they can actually further complicate the response.[242] Getting donations to the affected community can be a logistically challenging, costly, and time-consuming endeavor. Sometimes arrangements are made ahead of time and a local organization expects and is ready to accept the donations. This is helpful but at other times, donations arrive in the affected community unsolicited and with nowhere to go, which can slow the overall response. After Hurricane Mitch, for example, planes carrying aid to Honduras were unable to land on the runway because it was covered with piles of donated clothing that no one had the capacity to manage.[243]

Assuming donated items do successfully arrive in the affected community, are timely and appropriate, and there is someplace to put them, someone has to organize and distribute them—which is what this volunteer in Texas was doing. Surely there were evacuees who could benefit from the donated clothing, but how would they know it was at the shelter? As it was, I had been there for hours without anyone mentioning the donations upstairs. Further, judging by the fact that only a single volunteer was managing the donations, I suspected others at the shelter were also unaware.

Once the response ends, someone has to figure out what to do with the leftover donations. It is not uncommon to hear about disaster donations filling entire warehouses that are never used. In situations like these, the best-case scenario is that the donations sit in the warehouse for a few months until another organization agrees to take them. Often, though, there is just too much, and communities have to dispose of the donations. Following the Indian Ocean tsunami at least one community in Indonesia was so overwhelmed with clothing donations that they piled them on the beach, and when they started to rot, burned them.[244]

In Houston, while the donations to the kitchen seemed to suit their needs, other types of donations—like these overwhelming piles of clothes—did not. Months later I tried to figure out what had happened to the leftover clothing, but no one seemed to know. I suspect someone just threw it all out.

I did find places, though, where donations were successfully being handed out. One church in Houston had become a hub for assistance. A few national disaster nonprofits had set up shop in the parking lot and were handing out hot meals and information about how to begin recovery. The church itself had transformed into a donations distribution center, run and staffed by church members. Their location at the intersection of flooded neighborhoods made them easy to find.

Residents who were affected by the flooding drove up, provided identification, and then volunteers filled their cars with cleaning supplies, nonperishable food, diapers, and other essentials. The donations being distributed seemed to match the needs of the people who showed up looking for help. One volunteer told me they had collected cash donations when the flooding started and purchased the supplies they were now handing out. They were benefiting from the strategy that most disaster experts recommend—donate cash rather than things, so people on the ground can go buy exactly what they need.

As I stood watching the long line of cars drive up, I wondered about everyone who was not there. Where were the people who did not have a car, or whose car had been flooded? What about the people who had a physical limitation that prevented them from coming to the church? I understood why, from the church's perspective, they wanted to check IDs—they had an obligation to their donors to make sure the donations were going to help those affected by the flooding—but what about the people who needed help and did not have an ID?

This was something several volunteers talked about throughout the week.[245] Undocumented residents who had been affected by the flooding needed help, but they were understandably afraid to stay at the shelters with police standing outside, and had no government-issued identification. Volunteers slid them as many donations as they could and tried to direct them to organizations where long-term assistance might be found. None of the volunteers seemed confident, though, that they had given out the right advice.

I heard similar stories and saw similar issues everywhere I went, from the middle of Houston, to the suburbs, to the rural counties. In each of these places I found local spontaneous volunteers, often working alongside local nonprofits, running shelters, making meals, distributing donations, and coordinating their own response.

As the water cleared out, the city moved the people still staying at the shelters into hotels. The start of the long road to recovery had begun, which was my signal that it was time to leave. I spent my last night in Houston at a hotel packing up for an early flight back to Fargo. I had Beyoncé's recently released visual album, *Lemonade*, playing in the background. "Formation" came on and I watched as Beyoncé lay across the top of a submerged police car in the middle of a flooded New Orleans.

I fell asleep and a few hours later was woken up by a knock on the door and commotion in the hallway. A tornado had been sighted nearby, and we were under a tornado warning. I put my shoes on, grabbed my pillow, and went out into the more sheltered hallway. Though a bit disoriented, I chatted with the woman from the next room over. Her apartment had flooded and they had just moved her into the hotel from one of the shelters.

Up to Our Bottoms in Alligators

IN TEXAS, I had spent much of the week in my rental car, living off of granola bars as I drove from one flooded community to the next. At night I narrated my observations into a voice recorder, too tired to actually write them down, and then passed out on my bed still fully clothed. It had been a whirlwind. On the plane back to North Dakota, I finally had a moment to catch my breath and start to process the bigger picture of what I had seen. Although I had been there to study the volunteers, I also talked to emergency management officials, other community members, and survivors. Everyone had one thing in common—they were tired.

What was remarkable to me about the Tax Day Flood was that the response was so unremarkable. I would describe what unfolded in Houston and the surrounding communities as an average, run-of-the-mill disaster. It was not unexpected or even unusual. There was no controversy, no colossal screwup that left hundreds dead. There were no scandals, or paper towels thrown. The response mechanisms that emergency management had planned for seemed, for the most part, to be functioning.

Bigger than an emergency but smaller than a catastrophe, it was large enough to require a federal response, but not so large that a slow federal response would have been felt on the ground. While it does not feel right to call the response a failure, there are some in Houston—like the family on the mattress—who probably would not call it a success either. People were ultimately rescued from their homes and evacuated to shelters that had food to serve but parts of the response could have been done more efficiently and effectively.

Where needs were going unmet, volunteers were there to help. While the media often tells the stories of regular people spontaneously jumping into their boats to go save their neighbors, local churches opening up their doors, and local chefs grilling burgers in the middle of the street to feed volunteers as though they are unique moments of heroism, they are actually common and predictable.

As an outsider, these unplanned efforts can appear chaotic. At times, the involvement of volunteers is interpreted as a failure of the government's response.[246] This was certainly felt to be the case among many spontaneous volunteers I spoke with who had found themselves in unexpected leadership positions running shelters. There can be tension between these volunteer efforts and the government's response as they are unaware of what the other is doing. This tension was certainly there in Houston. Over and over again I heard from volunteers that there had been no plans for opening shelters, yet extensive sheltering plans had been developed by the Red Cross and local emergency management officials. Some of the plans were even available publicly online. This kind of disconnect is not only frustrating for all involved, but it can also complicate the response.

I thought about how disaster researchers had at times described the amount of help that arrives at the scene of a disaster as a "mass assault." I did not see a shortage of volunteers during the response in Texas, but I also did not see extra volunteers just hanging around with nothing to do. For all the volunteers who were there helping, even running various aspects of the response, there wasn't anything approaching what I would call a "mass assault." For a disaster that covered such an expansive geographic area it felt like I could fit the response in my pocket.

I left Houston worried. The Tax Day Flood is not the kind of disaster Hollywood makes movies about, but it is representative of the kind of disasters communities regularly face. There will always be catastrophic events like Maria and Katrina and large-scale disasters like Harvey and Sandy, but it is the repetitive, smaller floods that communities are being forced to navigate more frequently that seem to be wearing us down. East Texas had been through three floods in the span of just thirteen months. Less than two months after I left, Houston would flood again. Some people reflooded while some escaped the next round. Still others who had not previously flooded did so for the first time. It was a kick while they were down. Yet everyone was doing what they could to keep moving forward. There wasn't any other option.

Four months later, in August 2016, I returned to Houston to interview volunteers still working on the recovery from the Tax Day Flood, and I found the fatigue was even more dire. I was walking into a region that, though well practiced in responding to floods, was tired and low on resources. At one community meeting I attended, a woman raised her hand and explained that her home had flooded multiple times in recent years and she couldn't do it anymore. She wanted a buy-out. The mayor breezed past her question. Houston was clearly trapped in a cycle of recovery.

During my recovery trip I asked a number of volunteer coordinators and people working at the national disaster nonprofits how they felt their recovery efforts were going. The people who ran these volunteer organizations were proud of the work they were doing but were running into problems. They were low on money and struggling to find enough volunteers to finish the homes they were working on. They all raised concerns about "volunteer fatigue" and "donor fatigue" citing the number of disasters that had occurred across the country. Their organizations had stretched themselves thin trying to be in multiple places at once.[247]

These problems weren't just happening in Houston; I heard the same story everywhere I went in Texas. In fact, there seemed to be even less help in the rural towns outside the city. I spent one afternoon driving around a recently flooded town with their emergency manager. He at first struck me as the stereotype of a Texan emergency manager—gruff and tough—but as we drove from street to street I also saw that he was defeated.

We saw neighborhood after neighborhood with dumpsters lining the streets and contractor signs posted in front yards. The scene transported me back to post-Katrina New Orleans. There were people outside working on their own homes. I noticed camping tents had been set up in their front yards. The emergency manager shook his head and said some had tried sleeping inside their flooded home for a while but the mold was so bad they finally moved outside to the tents.

He told me about the work some local nonprofits were doing. It was immensely helpful, but it also wasn't enough. A few weeks earlier he had finally been able to get into contact with one of the national nonprofits. They sent out volunteers but could only commit to one weekend of help. As we drove he told me the story of the response and described the process of trying to get state resources in to help. The Red Cross never showed up so the local church had opened the shelter for people who needed a place to stay.

As we drove he kept the window down, despite the heat, so that he could wave and check in on everyone we passed. He told me about the mitigation plans the town had created almost two decades earlier. The plans were just sitting on a shelf in his office because they hadn't been able to find any federal funding. At the end of our day he summed up his town's flood situation as only a Texan could: "We're up to our bottoms in alligators."

PART FOUR

Damage Control

"We have so many disasters bombarding us right now, my dear."

—CATHERINE O'HARA, *SCHITT'S CREEK*

DRIVING WEST FROM New Orleans, you pass into the suburbs of Metairie and Kenner and then turn south through the
rural towns of Des Allemands and Lockport. By the time you
reach the little, aging town of Larose, you start to wonder if this
road will take you to the end of the world. For the remaining
fifty-mile stretch to the coast, the scenery passes like carefully
curated displays of the causes and consequences of our changing
world. You are flanked by disappearing wetlands—the seas are
rising, the land is sinking, and the shores are eroding.

It takes a moment, on your first drive down Highway 1, to
realize that the telephone poles were not intended to be away
from the road in open water. There used to be land there. It
takes another moment to realize the roadside docks are not for
recreation, but rather to transport owners to their homes, which
are a hundred yards back from the road, raised a ludicrous fifteen feet in the air as water flows freely underneath.

Highway 1 runs out of land in a little sliver of a town, Grand
Isle, a barrier island with one main road.[248] Houses, most on

stilts, line either side of the seven-mile-long island. There's a lime-green snowball stand, a couple bars, and tourists in the warmest months. A defiant bunch, the residents of Grand Isle, Louisiana, have seen twelve-foot storm surges, one-hundred-mph winds, and had four million barrels of oil washed up on their shores. Some have moved away, and most who remain have rebuilt again and again. Grand Isle exists at the nexus of the climate crisis: the oil and gas industry, our built environment, and nature.

It was not until I spent time in southeast Louisiana during the BP Oil response that I fully grasped climate change was not just a challenge for the future, but for the present as well. As I've seen the mark of the climate crisis across Louisiana, along the coast of Maine, and in Texas, I've come to understand it is everywhere, and we're not ready.

Climate Change Is Already Here

I GREW UP a cold-weather girl who went sledding without a jacket and went for midnight winter beach dips in the Atlantic. The heat in New Orleans took my breath away and the humidity made me feel like my skin could melt. In New Orleans we built houses under the Southern sun. We worked through the heat, guzzling water, bathing in sunblock, and seeking shade in tool sheds. A few months after the Tax Day Flood, when I returned to Houston to interview recovery volunteers, I found nonprofits calling their workdays short. The heat index was around 120 degrees and it was too dangerous for their volunteers to be outside gutting flooded homes. They hoped the extreme heat would settle soon so the weather wouldn't further delay the recovery efforts.

The world is warming fast. The global average temperature has risen 1.8°F since 1900.[249] It doesn't seem like such a small number could matter that much, and yet rising temperatures are increasing the frequency and intensity of heat waves around the world.[250] In the United States the annual average temperature is expected to increase by 2.5°F by 2050.[251] We are talking about the kind of heat where drinking an extra bottle of water and taking five-minute breaks in the shade isn't enough.

In the summer of 2003 a heat wave blanketed Europe. The warming forced the evacuation of people from the Matterhorn when melting triggered a rockfall, trains in London were shut down, and forest fires cut through parts of Spain, Portugal, and Italy.[252] Human-induced climate change doubled the likelihood of the 2003 heat wave.[253] Over 70,000 deaths were attributed to heat-related complications.[254] It is difficult to conclude precisely how many of those deaths are attributable to climate change, but one study found that of the 753 heat wave–related deaths in Paris 506 could be linked to climate change.[255]

Blistering temperatures have circled the globe in the past two decades, including in the United States. Phoenix is on the front lines of these warming temperatures. The city has seen an increase in days above one hundred degrees and evening temperatures are on average nine degrees warmer than they were in 1948.[256] The temperatures are so extreme that in the summer the city becomes part nocturnal—going for hikes and doing road construction once the sun goes down.[257] Like in Europe we are breaking records for heat-related deaths in the US. A 2019 heat wave in Arizona killed 187, breaking the state's heat-related deaths record. Heat waves have had a disproportionate effect on the homeless, low income people, and the elderly among other marginalized groups.[258] In Phoenix, researchers have found that while half of the warming in the city is attributable to climate change, the rest is the result of an "urban heat island effect," caused by how we have built our cities (e.g., asphalt parking lots and paved roads, building sprawl, cars, air conditioners).[259]

Our cities haven't been built to absorb this kind of heat.[260] Places that have historically had longer periods of sustained heat may be more prepared because their office buildings have central air and homes were designed to encourage air circulation. Places that have not traditionally dealt with this kind of sustained heat will struggle. I've already noticed the change in Maine. In recent summers, as temperatures rise, Mainers have rushed out to buy AC units to stay cool.[261] It's not surprising given homes in the Northeast have traditionally been built to keep the heat in.

There are photos of me throughout my childhood standing hidden within a snowsuit standing in front of an enormous pile of snow. Maine's snowfall totals are impressive. On the southern coast the average annual snowfall is around six feet, and in the north the average is around eight feet.[262] Although it is certainly an endeavor to shovel, the snow provides vital nutrients for the soil, replenishes our drinking water, protects against drought, and is a key driver of the local winter economy. But this winter weather can also be dangerous.

In January 1998, everything in New England stopped as our world disappeared under three inches of ice.[263] The state went dark when as many as three million feet of power lines were knocked out. Normally my family would have just hunkered down by the fireplace under a sea of blankets but my brother had been born prematurely just a few months earlier. Living without power, or heat, wasn't an option this time. My parents moved us into a hotel, seemingly the only place left in the state with power. I was seven years old and took up residence at the indoor pool while outside cars slid off the roads, thousands of others stayed at emergency shelters, and hospitals treated patients for hypothermia, carbon monoxide poisoning, and falls on the ice. When we were able to return home, we found tree branches blanketing every inch of our yard. I helped my dad clear them as the sounds of chainsaws filled the neighborhood.

Afterward there were "I survived the Ice Storm of '98" T-shirts made, which to me implied the experience was a novelty. I thought it would be a once-every-few-decades kind of disaster. Despite what the term "global warming" implies, climate change is also impacting cold weather and winter storms.

Although average temperatures are increasing when it is cold, we can expect winter storms to be more intense than they have been historically. Nearly twice as many extreme snowstorms occurred in the second half of the twentieth century as compared to the first half.[264]

As much damage as the ice storm of '98 caused in Maine, when this kind of weather hits more populated states it can lead to even bigger problems. In 1993, "The Storm of the Century" brought the country to a standstill as every major airport on the East Coast shut down.[265] Almost three hundred deaths were attributable to the storm. An increase in traffic accidents threatened to overwhelm hospitals and push communities' resources to the brink. Homes collapsed under the weight of snow. In all the storm caused upward of $9 billion in damage.

As is the case with rising temperatures, some places are better equipped to handle these changes to winter weather. New England cities will amend their snow removal plans to handle more intense storms. It will require them to make some changes and may be expensive, but they'll probably be able to manage it. Winter storms are well ingrained in the culture and their towns have been built for it. Other places, though, will require more significant adaptation measures.

In 2014, Atlanta succumbed to a snowstorm, dubbed "Snowmageddon," which brought about two inches of snow and ice to the southern city.[266] These exact same conditions in Boston would have been just another Tuesday, but in Atlanta it brought the city to a stop. Hundreds of accidents occurred throughout the city as people tried to get home on icy roads. Some children stayed at school overnight while other commuters had to take refuge in nearby stores and businesses that opened their doors. Atlanta didn't have the equipment or systems in place to manage this kind of storm and so the impacts were more severe.

Temperature changes and shifting snowfall totals are beginning to affect spring floods. The 2018-2019 winter saw unusually high snowfall. Then in the spring, warm temperatures and significant rainfall hit a still frozen ground that could not absorb the melting snow.[267] Everything flooded.[268]

I spent the spring of 2019 monitoring the height of the Red River located just a few blocks away from my apartment. The Red River had last seriously threatened downtown Fargo in 2009.[269] In the following years, the city did buy-outs, changed drainage, put parks in next to the river, installed floodgates on the bridges and more all in an effort to prevent flooding.[270] Despite these mitigation efforts, by March fliers were hung up at Starbucks asking for volunteers to help fill sandbags. The mayor wanted to fill one million bags, just in case.[271] I spent an afternoon at Sandbag Central in an assembly line tying off bags before they were stacked and carted away.

Across the river in Minnesota, the streets closest to the river started to flood. I climbed up to the roof of a parking garage and looked out at the river encroaching on Fargo. The city was ultimately spared this time, but I didn't know how much longer the mitigation efforts would be able to keep up. I drove north of the city and found backyards and garden sheds underwater. Homeowners had put up temporary barriers as a final line of defense. I passed by barricaded side roads and saw trains barreling down railroad tracks that remained just a few inches above water. I drove down the endless rural county roads as the sun reflected off the glassy smooth pools of water that covered acres of farmland.

The flooding in North Dakota was just part of the extensive flooding in the Midwest that year. Rivers throughout the middle of the country began to overflow and the Mississippi River broke the record for the longest continuous flood since 1927.[272] Fourteen million people were affected, and eleven states requested Presidential Disaster Declarations as they became overwhelmed by the damage.[273] The National Oceanic and Atmospheric Administration (NOAA) estimated $6.2 billion in losses.[274]

That August I moved from Fargo to Omaha, Nebraska, and found the same conditions.[275] Flooding across the Missouri River basin had breached levees and washed bridges away. Highway exits and parks disappeared under standing water for weeks, and in some cases months. I moved into a new apartment just one street away from the river and continued my habit of monitoring river heights. I drove out of the city, to the small towns and rural communities that were trying to rebuild. I met with emergency managers who were trying to get the mitigation funding to prevent their communities from flooding again and eavesdropped on locals sitting in coffee shops debating what the state would do. I told you—same story, different disaster.

Changes in snowpack and snowfall that contribute to flooding can also have the opposite effect. Droughts are increasing in intensity in some regions of the United States.[276] Shrinking snowpack and changing melting times is threatening not only flash flooding but also the drinking water supplies of millions.[277]

Droughts as consequence of climate change are of particular concern because they can put pressure on already struggling food and water systems. If not managed effectively, water shortages can lead to food insecurity or even famine, which can, in turn, contribute to the conditions for increased conflict. Civil conflict can lead to mass migration as millions become displaced internally and many more millions seek refuge in other nations. This can strain legal and healthcare systems, and create further instability that spills out into neighboring countries. Climate change is far from the only factor at play in these types of situations but, as the US military calls it, climate change is a "threat multiplier."[278]

The changes in expected drought conditions and hotter weather have also changed the predictability, intensity, and frequency of wildfires.[279] Since the 1980s, wildfires in the western US have grown in number and severity,[280] and we are now breaking fire records so fast that whatever statistics I write here will be outdated by the time you read them.

I was in the middle of a road trip with a friend who was moving across the country the first time I saw how quickly these fires can spring up. We were driving through narrow canyon roads outside Sedona. In what felt like just a minute the air started to hang thick and we could smell burning. A flashing road sign advised us to turn on the radio for warnings and evacuation orders. Street names and neighborhoods blared out of the radio at us but with no cell service we had no idea where we were in relation. The fire we were near ended up being brought under control and we were never in any real danger, but as I had felt in the days before Gustav, I knew I had been vulnerable.

Finding ways to effectively communicate quickly changing warnings to the public is becoming more challenging and urgent as these fires worsen and the fire season lasts longer. Each year the wildfire season out West starts earlier and ends later. In recent years the wildfire season in California is now as much as seventy-five days longer than it used to be[281] and some are advocating calling it a "fire year" rather than "fire season."[282] As one wildland firefighter put it, "there are no climate denialists on the fire lines."[283]

Wildland firefighting is an unforgiving, brutal job that wears at the body and on the soul. It also requires extensive resources. Moving forward we need to quickly grow our firefighting capacity to meet these growing needs. This is an issue internationally as well. The United States and Canada have mutual aid agreements with Australia and New Zealand. Historically, our fire seasons have been opposite to one another, which allowed us to lend each other support—US firefighters would go to Australia and Australian firefighters would come to the US. In recent years because of the lengthening fire seasons and the severity of those seasons in both the US and Australia the ability to fulfill those agreements has become more difficult. These are the types of approaches that are going to need to be rethought as these changes continue.

It's not only the fires themselves we need to contend with but also the flash floods and landslides that can come after.[284] As the frequency and size of fires increases so too do these types of threats.[285] Following the Thomas Fire in 2018 a debris flow through Montecito, California, left twenty-three dead and hundreds of homes destroyed.[286]

There's another link between climate change and landslides too. Researchers working on the Barry Arm Glacier in Alaska recently discovered a significant landslide-generated tsunami threat. As the climate changes the Barry Arm Glacier has been rapidly retreating, leaving behind unstable terrain. If this land suddenly failed it would generate a tsunami and waves as high as thirty feet could reach Whittier, Alaska.[287] As glaciers continue to retreat and permafrost continues to melt there is an increased risk of rockfalls and landslides.[288]

I had never been in a real rainstorm until I lived in New Orleans. The sky darkens faster than you can run inside and the thunder shakes so hard it feels like the city will snap in two. A wall of water comes from the sky and ricochets violently onto the cracked pavement and concrete. The pumping system, designed in the early twentieth century, can't keep up with twenty-first century rains, so the streets fill with water.[289] I'd run through knee-deep water, holding my shoes, on my way back from moving my car to the neutral ground—to prevent it from flooding. The ground was invisible beneath the water and it was nothing but luck to make it back inside without stepping on glass, falling into a concealed pothole, or worse. Rinse and repeat.

It doesn't take a hurricane to flood the streets of New Orleans. Changing temperatures influence moisture in the atmosphere, which has led to more frequent and intense rainfall.[290] In the United States, the Northeast has seen the greatest increase so far, but other areas have been affected too.[291] The Tax Day

Flood in Texas, the 2016 Louisiana flood in Baton Rouge and surrounding areas,[292] and repetitive flooding in Ellicott City, Maryland,[293] are just a few of the recent disasters that have been spurred on simply by heavy rainfall. Our infrastructure wasn't built to handle so much water, so fast—so it floods.

Climate change affects the amount of rainfall hurricanes bring too. The flooding from Hurricane Harvey in 2017 and Hurricane Florence in 2018 were largely a result of their exceptional rainfall totals. Both storms crawled along slowly, giving rain more time to fall. Harvey's rainfall totals were 10 to 38 percent higher than would have been expected in a world without climate change[294] and Florence, a year later, was found to have 10 percent more rainfall.[295] Warm water is hurricane fuel, and so as climate change warms our oceans, there is more opportunity for storms to intensify. Researchers have found a link between climate change and the increasing trend of Atlantic hurricanes rapidly intensifying (when maximum sustained winds increase quickly).[296] The 2020 Atlantic hurricane season demonstrated why this is such a concern as one storm after another underwent rapid intensification right before landfall. Rapid intensification is a huge challenge as there may be less time to issue appropriate warnings, evacuate, and take other protective actions.

In addition to flooding from rainfall, hurricanes also bring storm surge (the abnormal rise in seawater because of a storm).[297] In 2012, Sandy hit New York and New Jersey with a nine-foot storm surge that coincided with high tide and sent a record-breaking high tide of fourteen feet onshore. The surge inundated homes, businesses, and the city's underground infrastructure. If the same exact storm had occurred at the beginning of the twentieth century, the storm surge would have been over a foot lower. In Lower Manhattan, fifteen inches of sea level rise and subsidence has been recorded.[298] A foot of difference might not seem like much. Thirteen feet of water still would have caused extensive damage but research in the years since has found that for every additional inch of sea level rise an additional six thousand people were affected during Sandy.[299] Even with as much damage as Sandy caused, it was far from the worst-case scenario for the East Coast. Higher tides and stronger storms are in the Northeast's future.

There are many places along the coast where it doesn't even take a storm for the roads to flood. Sea level rise is already leading to flooding at high tide—what some call "sunny day flooding"—all over the country. Places like Galveston, Annapolis, Charleston, and Miami are already battling this persistent flooding.[300] By 2030 the frequency of this type of flooding could as much as triple.[301] This can add up to serious impacts including impassible roads, flooded homes and businesses, and damaged infrastructure and systems.

In Isle de Jean Charles, Louisiana members of the Biloxi-Chitimacha-Choctaw and United Houma Nation tribes are relocating inland, as the land they live on is claimed by the Gulf of Mexico.[302] The land is subsiding, and the water is rising. They've named themselves America's first "climate refugees."[303] The last time I was in Isle de Jean Charles it was a clear day—for both sky and road. A few people remained, sitting out on their front porches, but many more houses had been boarded up. It won't be long now until the road to Isle de Jean Charles is permanently underwater.

Sometimes if you do not know what to look for, the fingerprints of climate change are hard to spot. When I first learned about climate change growing up, we were presented with what amounted to sci-fi images of the apocalypse, which is why I was surprised to first recognize climate change in the quiet calm of Louisiana's Highway 1. The damage from climate change is already all around us, though, seeping up from the ground below and falling from the sky above.

I saw climate change in the canals of Venice where I lay down in the middle of walkways to take photos of the city's drainage that struggles to manage increasing flooding.[304] I saw it in the fields of lupines that covered the mountainsides of Iceland, a sign that the climate had changed enough to let them grow.[305] I saw it in Budapest as I stood in the shadow of a relief on the side of a church commemorating the 1838 flood just a few blocks away from the Danube River, which had been flooding more frequently in recent years.[306] I saw it in the faces of the students I met with on the top of a mountain in Vazhayur, India, who told me about their volunteer efforts during worsening floods in Kerala.[307] I saw it in Colombia when a friend and I had to duck into a grocery store as a rainstorm submerged the streets of Cartagena and locals rushed to tell us not to worry, that this was normal now, and that it would pass.[308] I reassured them I had lived in New Orleans and understood.

I see sea level rise when I look at the ocean, fuel when I look at the forest, and carbon emissions charts on seventy-degree New England January days. I see climate change in the increasing number of billion-dollar disasters. I see it in the increasing disaster relief budgets. I see it in the disaster survivors sleeping in tents next to their half-gutted homes. Climate change is everywhere.

Scientists have developed a methodology called "extreme event attribution" that helps us understand the relationship between any given weather event and the changing climate.[309] This work is poised to be an empirical link between climate and disaster science. The culmination of over three hundred peer-reviewed studies have thus far painted a clear picture of how human-caused climate change is altering global hazard risk. Carbon Brief analyzed all extreme event attribution studies done through 2019 and found that of the events that have been studied 69 percent were found to be made either more severe or more likely because of climate change.[310] This research confirms we are living in a world with a changing landscape of extreme weather. It is a world where more people in more places are facing more frequent and severe threats, in part because of the changing climate.

Not every climate-fueled extreme weather event is a disaster. Hurricanes form out at sea and never reach the shore, and plenty of wildfires are put out before destroying a town. Rivers are meant to overflow banks to replenish the soil, and we need rain to water crops. But climate change means that the behavior we have come to expect from these types of threats has changed, and if we want to protect ourselves, we have to change, too. We have always had to manage our risk and we have always had disasters, but climate change, along with other realities of our current world, have become entangled to create unprecedented risk across the country.

As with any type of hazard, it is when these climate-related hazards threaten us, our environment, and our livelihoods that we begin to have a problem. It is when a hurricane makes landfall and destroys a city or when a wildfire burns so fast that there is no time for evacuations, that we have to manage the fallout. It is when managing these events overwhelms the capabilities of the affected community, that these events become disasters (or catastrophes).

Climate change is so insidious because it intertwines itself with our existing vulnerabilities and amplifies them. It further threatens what is already fragile and at risk. It takes these hazards that we have always had to manage—hurricanes, flooding, landslides, wildfires—and changes the way they manifest. These changes, interacting with other factors like population movement toward high-risk areas, poorly written and enforced regulations, social and economic inequality, decaying infrastructure, and poor development decisions, create our risk. In other words, climate change paired with these demographic, regulatory, and policy factors are literally a recipe for disaster.

We have to assess the ability of our communities to absorb

these changes. How many simultaneous emergencies can a given town handle before they are overwhelmed? How many fires can we battle at once? How many floods? How many simultaneous disasters does it take to send our national response capabilities into a crisis of their own? How many of these situations that might have only been an emergency instead become a disaster, or a disaster become a catastrophe, because of the added pressure of climate change?

These changes don't happen in isolation from each other. We don't just have to deal with sea level rise or river flooding or a changing wildfire season. We have to do it all, and all at once. It's why, when I hear politicians say the solution to climate change is to have people move, I have a hard time understanding where it is they think we're all going to go.

While the coasts begin flooding due to sea level rise, they will also be battered by more severe storms and intense hurricanes that bring increased rain and storm surge. These storms will also occur simultaneously with other climate-related consequences, like increases in epidemics, conflict, and food and water shortages. As we have learned during the COVID pandemic, disasters that occur in unstable conditions can amplify dangers and are more complicated to manage.

Disasters don't hit us one at a time. They do not wait in line to begin until the preceding disaster has ended, which means we do not go into every crisis at full capacity. Climate change makes these events more severe, frequent, and otherwise different than what we have become used to, which threatens to overwhelm our resources and existing approaches to emergency management. When crises occur simultaneously, the overarching response system can become strained. While a catastrophic event immediately overwhelms the system, a series of emergencies and disasters can have a similar effect. Our approach to emergency management relies on help converging in from nearby towns, states, and the federal government, but what I have started to see is that this system is already struggling.

I remember standing on the roof in the Lower Ninth Ward as oil gushed out into the Gulf of Mexico and wondering how we would manage two recoveries at once. I've watched Camp Ellis become trapped in a cycle of crisis, unable to do enough mitigation before the next flood comes. In Texas I heard the people who are on the front lines of disaster response describe the fatigue they were feeling as they raced from one end of the country to another. My experiences in these places were all giant red flags that we aren't ready for what's coming.

We have to urgently work toward limiting our carbon emissions, but at the same time we have to manage climate change's consequences. We can't be so busy thinking about the future that we miss what's happening right in front of us. We have to start doing more to control some of this damage. Our risk is continuing to grow and so far climate change is kicking our ass.

Facing the Consequences

THE LAST TIME I drove down Louisiana Highway 1 was the spring of 2019. I was there with a group of undergraduate students, mostly emergency management majors, on a trip to teach them about disasters in Louisiana. The students were primarily from small midwestern towns. Most had yet to experience a disaster, let alone a catastrophe like Katrina.

We took the students all around New Orleans—through each neighborhood, past the levees, and to talk to city officials. I told them the story of Katrina, an event they were too young to really remember, and explained why we could never let that happen again. We drove down the road to Grand Isle where they were confronted with the same disappearing wetlands, rising seas, sinking land, and eroding shores that taught me the urgent risk of our changing world. As we drove, I noted that unless the entire highway is raised, I probably wouldn't have the chance to make too many trips like this again.

★ ★ ★

It is not a question of *if* we will experience the consequences of climate change, but rather *how bad it will be*. This is a vitally important question for emergency management because there's a big difference between preparing for just one foot of sea level rise and preparing for six feet. Risk assessments have to be based on future projections, but it's hard to fully know which projections to rely on, especially on longer timescales.

The timing and extent of climate policy is a bit of a wild card in figuring out what exactly we need to do to prepare and adapt. The severity and timing of climate impacts will depend on how quickly, and to what extent, we act to lower emissions. We face a range of scenarios that vary based on how quickly and completely the world moves to curb carbon emissions, and other factors that can influence changes in the climate.[311]

The true worst-case scenarios are so horrific most scientists leave it up to science fiction writers to put them into words. In a controversial *New York Magazine* article in 2017 writer David Wallace-Wells noted, "we suffer from an incredible failure of imagination" in how we talk about the worst-case scenarios.[312] On this path, we stay our current course, and our carbon emissions continue to increase.[313] The United States, China, and other major polluter countries fail to participate in global climate agreements and curb emissions. Elsewhere, countries that are currently industrializing do so to the same extent and using the same methods Western nations have, leading to higher emission rates. Changes cascade in the next few decades and the environment of *Mad Max* becomes a new reality for many.

Then there are the middle-of-the-road options. This is a world in which it takes us a long time to break up with fossil fuels. We do eventually make the switch to renewable energy, but not before the seas rise up to permanently flood coastal cities around the world. In many places the air becomes hard to breathe. Water and food systems become strained and even overwhelmed, and civil conflict breaks out around the globe. We have to manage an endless barrage of catastrophic events across the country and world.

A more optimistic scenario assumes we act in line with the parameters set forth in the Paris Agreement.[314] Globally, we drastically reduce our carbon emissions and switch to renewable energy. In this best-case scenario, we will *still* experience sea level rise, stronger storms, and hotter temperatures for decades.[315] Without significant adaptations, we contend with one disaster after another around the country. Many die from various climate-related crises like disasters, violent conflict, and other major displacement events. Sea levels rise significantly and if we do not adapt quickly enough, we still risk losing places like Miami, southeast Louisiana, and coastal towns from Alaska to Maine.

What this means is that no matter what road we go down—even our absolute best-case scenario, we will continue to experience disruption from climate change and will have to grapple with the fallout. People will still lose their lives and livelihoods, systems will be strained, and communities destroyed. There is no scenario that doesn't require adapting our communities. The question is, how much will we need to adapt?

We are well past the point of preventing human-induced climate change altogether, but there is still a long way between where we are now and David Wallace-Wells's description of "an uninhabitable earth." I don't have a crystal ball to tell you if the US and the global community are going to get it together in time to keep us in the range of the lower-end scenarios. So while there absolutely is the possibility, and hope, that we take effective action on climate change, the reality is that there is no way to know for sure that we will. Even if we do, we will be locked into a certain amount of change, and climate-related disasters will continue to occur.

Others can write about the missed opportunities and the turning points that led us to this reality. Blame, in unequal measures, will be placed among the many who could have done more—or at least something. None of that history changes our current reality. None of this reckoning changes our future. What we can still change is the severity of climate consequences and how we manage them. We need to find a way to live in our new world. We need to imagine what we want our communities to look like ten, twenty, thirty years from now. What does a safe community look like? What do we view as an acceptable amount of risk? How many floods? How many houses destroyed? How many lives lost? Assuming the answers to those questions are "as few as possible," then we need to make changes—now.

Elected officials need to understand this too, because we need them to make policy changes to enable emergency management to work effectively. In the early days of the 2020 Democratic primary, the Democratic National Committee (DNC) refused to allow candidates to hold a debate on the climate crisis. CNN did a poll in the summer of 2019 that found 95 percent of Democrats agreed climate change was a serious threat that nominees must address. Constituents protested, the candidates agreed, and yet it was forbidden. Tom Perez, the chair of the DNC, argued that if they allowed a single-issue debate on climate change, they would have to do the same for healthcare, immigration, foreign policy, and all the other issues constituents care about. What he didn't seem to understand is that the climate crisis is not *an* issue, it is *the* issue. It affects every other issue we all care about from healthcare to immigration.

Fortunately, CNN snuck around the DNC's decision by hosting a "town hall," not a debate, on the climate crisis. This is how I found myself sitting on my couch for *seven hours* watching the top ten Democratic presidential candidates take to the stage and answer questions from CNN hosts and audience members. It was a marathon. For the first time in history, CNN had dedicated itself to climate change. The purpose of the town hall was to get an understanding for how each of the candidates understood climate change and what they planned to do about it. Candidates rushed to release their climate plans with some coming out just hours before the town hall, reminiscent of a college student pulling an all-nighter to finish their final paper that they hadn't started until the night before it was due.

This appropriately sums up the climate crisis. Our leaders have procrastinated. Now, the assignment is due and everyone is turning in whatever they've managed to get done in the final hours. It's that last-minute, pulled together overnight, no time to hit spell-check kind of thinking that we professors recognize with ease. Here, the consequence is not a failing grade, it's a failing planet.

Most of the candidates' climate plans didn't meet the urgency of the moment, but most notable was how few candidates mentioned anything about emergency management policy. If we know that climate-related disasters are already increasing, and will continue to increase into the future regardless of what climate action we take, the only responsible response is to prepare our communities—and emergency management—for those disasters. Yet, while I watched these town halls this was hardly mentioned.

One of these town halls took place as Hurricane Dorian was hovering over the Bahamas causing catastrophic damage.[316] The satellite image of the hurricane was displayed as a literal backdrop to the event. As members of the audience asked questions it was revealed that many of them were themselves disaster survivors. To the candidates' credit they did all agree on some fundamental truths—climate change is happening, the impacts of the consequences are unequal, and we need to do more to prevent disasters from happening.[317] Things went off the rails from there though with most candidates unable to articulate effective policy changes for addressing these issues.[318] If the low bar is politicians who think climate change a hoax, then this group of candidates were doing well, but I also was left unconvinced that the candidates fully understood the colossal mess that we are in. We deserve more.

The Democratic primary illustrated the disconnect that I often hear in the way the climate crisis is discussed. The consequences of climate change—often disasters—are used as the basis for the argument that we must act to stem emissions. Pictures and video of devastating disasters and disaster survivors are used to justify climate action, but then those same people brush aside discussions of how emergency management needs to change. The climate conversation has to this point rarely included, or even acknowledged, the actual needs of the people who were affected by the very disasters they purport to be concerned about. It feels like a vestige of the old idea that climate change is *just* a future problem, rather than a problem for us now. They need to grasp that we not only need to reduce emissions, but also have to adapt.

I do not know which of these scenarios will be closest to our future, but I do know that regardless of where we end up, emergency management will be on the front lines of managing the consequences. Emergency management has always been important, but the climate crisis raises the stakes. We could continue using the system that we have. It will probably continue to trudge along and manage the smaller disasters we face relatively well. It will likely even be able to give some high-risk communities a lifeline for a few extra years. The limitations of the emergency management system will become even more visible exactly when the resources to fix them will become scarcer. Soon enough it will become constantly overwhelmed—unless we make changes.

In emergency management we constantly evaluate changing risk and, ideally, use those risk assessments to drive our decision-making. We need to factor these climate scenarios into those assessments so that our system is ready for whichever scenario ends up being our reality. We also have to do this immediately because, as of now, our emergency management system is not prepared to handle any of these scenarios—a truth made visible in 2017.

"A Real Catastrophe Like Katrina"

ON SEPTEMBER 20, 2017 nearly everything was working against Puerto Rico, including a history of US colonialism, a racist administration, a toxic political climate, a disaster-fatigued country, an overwhelmed FEMA, a brutally incompetent government, climate change, and a giant hurricane. An estimated five thousand Puerto Ricans were killed and hundreds of thousands more were left without help in the largest US catastrophe since Hurricane Katrina and the levee failure.

Decades of US policy and local politics laid the groundwork for the catastrophe that was Maria. In Puerto Rico, the colonial legacy and subsequent complicated legal status of the island left them particularly vulnerable. Decisions made in Congress, as well as within the Puerto Rican government, have stifled economic growth and independence. Neoliberal policies, including privatization and austerity measures, have created their own perfect storm of suffocating debt and widespread poverty.[319]

Before Maria, Puerto Rico was contending with a financial

crisis.[320] Unemployment had been on the rise since incentives for corporations to move to the island ended in 2006. Between decisions made in Congress and on Wall Street, Puerto Rico was $70 billion in debt before Maria.[321] As of 2013, a distressing 45 percent of residents[322] were living in poverty, including 60 percent of children.[323]

Not everyone is equal in the face of a hurricane—disasters *do* discriminate. Race, class, and gender all influence how we are able to prepare, protect, respond to and recover from disasters.[324] Someone who is in debt or living paycheck to paycheck must focus on their most immediate needs, leaving them unable to plan for future catastrophes. If someone does not have money to buy their kid new shoes, they do not have the funds to build a sturdier home, stockpile emergency supplies, or buy extra insurance. Similarly, local governments that do not have the money to maintain basic infrastructure also do not have the money to harden infrastructure to withstand major hurricanes. In the days before Maria made landfall, there was not much that could be done to reduce the longstanding social and physical vulnerability of everything from the electric grid to people's homes. You cannot fix decades (to say nothing of centuries) of bad policy in the hours before a hurricane makes landfall.

At the risk of stating the obvious, Hurricane Maria was a big and intense storm deemed "an absolute monster" by the *New York Times*.[325] Maria made landfall in Puerto Rico as a Category 4 with 155-mph winds, and a six- to nine-foot storm surge. As Maria moved across Puerto Rico, it dropped thirty-seven inches of rain—the equivalent of a quarter of Puerto Rico's annual rainfall.[326] The rainfall triggered landslides in 75 percent of the island's municipalities.[327] Climate change's mark was visible in Hurricane Maria. One study found that Puerto Rico was five times as likely to experience the amount of rainfall received during Maria in 2017 as compared to the 1950s.[328] On top of Maria's strength, Puerto Rico had already sustained damage during Hurricane Irma only weeks earlier.[329] While storm size alone may not cause a disaster, it would have been difficult for any place to contend with the sheer size and strength of Maria.

Hurricane Maria destroyed communications infrastructure across Puerto Rico, so it took days for many people on the mainland US to understand a catastrophe had occurred. It took many more days, and in some cases weeks, to reach the remote villages to learn the full extent of the destruction.[330] Nearly eight hundred thousand homes were damaged or destroyed.[331] Nearly everyone on the island lost electricity, water, and cell service. It would take $3 billion and eleven months for most power restoration to be completed.[332] The main seaports and airports were damaged, which contributed to fuel and food shortages across the island. When supplies did finally begin to arrive in San Juan, distribution across the island was difficult as debris blocked roadways.

Meanwhile the president lashed out at local politicians and claimed Puerto Ricans "want everything to be done for them."[333] This, of course, was far from reality. Puerto Ricans were organizing relief efforts, as survivors always do. The problem was that there were so many urgent needs that the efforts of survivors were not enough. As any community in their position would be, they were overwhelmed and needed help to come from outside—a defining characteristic of a catastrophe.

The problem was that many on "the outside" didn't seem to understand the extent of the crisis. The media had only weeks before done a fine job covering Harvey and Irma. Sure, Chris Cuomo standing on the beach scolding tourists for not evacuating felt closer at times to entertainment than reporting, but they were at least physically present. The national media were sharing important life-saving information, holding people accountable, explaining how to help, and keeping the public informed of the unfolding situation. In other words, what the media is *supposed* to do during disaster. For days, however, there was barely any mainstream coverage of Puerto Rico in the United States.

It was immediately clear to disaster experts that Maria would surpass the size of Katrina by almost all metrics. For fourteen years the media had obsessively compared every crisis to Katrina—headlines claimed everything from BP to the sloppy rollout of the Affordable Care Act website were "Obama's Katrina." Reporters had breathlessly weaved Katrina into Sandy coverage and just weeks earlier done the same with Harvey. For the first time, a comparison to Katrina was actually appropriate, and yet, as Maria unleashed hell across Puerto Rico, most of the national media barely mentioned it. After Maria, FiveThirtyEight[334] and the *Washington Post*[335] analyzed the coverage and their findings confirm the widespread media failures.

Only five hundred news outlets published stories about Maria compared to eleven hundred that covered both Harvey and Irma weeks earlier. Three times as many stories were published about Hurricane Harvey than were published about Maria. On the first Sunday after landfall, the network talk shows dedicated less than one minute to Puerto Rico *combined* as the island went without power, communication, and necessary medical care. In fact, it was only when the president started a public fight with the mayor of San Juan, Carmen Yulín Cruz, that media outlets increased their coverage. The disparities in coverage are particularly heinous when you consider Puerto Rico's situation was more dire than either Texas's or Florida's had ever been.

The media is one of the most important participants in emergency management. Survivors depend on it for information about the disaster itself—if they need to evacuate, how to stay safe, where to find shelter or food, to tell their story, and to communicate to the outside world what is happening and what they need. Those outside the disaster depend on media to tell us what is happening, if our family and friends are safe, and how we can help. Furthermore, the media is supposed to hold emergency management accountable. When help doesn't arrive and government officials fail to respond effectively, it is the responsibility of media to illuminate their failures. Accurate, appropriate, and timely disaster coverage literally saves lives.[336]

In the absence of robust national media coverage, regular people tried to fill the gap through social media. Yet, even they could not fully capture the unfolding catastrophe in Puerto Rico at first. For days, only 5 percent of the population had cell reception.[337] This stood in particularly stark contrast to Texas, where throughout Harvey, only 4 percent of cell towers went out during the storm.[338] Most Texans were able to maintain communication with neighbors, friends, family, and response organizations. They were able to post updates regularly, ask for help through social media, and keep up with life-saving news reports throughout the storm.

In spite of national media failures, local journalists in Puerto Rico were nothing short of heroic. They worked around the clock with the resources they had while many staff were unaccounted for or displaced.[339] Cell service and internet access were infrequent, and even when they had stories to tell, being able to disseminate them was difficult. Some journalists and their families moved into newsrooms where they worked continuously. Radio broadcasts were the primary source of news across the island as local journalists worked to get the word out.

Disaster response can be hampered by a lack of accurate coverage about the affected community's needs.[340] Media coverage can be a driver of disaster donations, as most donations are made within the first six weeks following the disaster.[341] In the month following Hurricane Harvey, the Red Cross raised $350 million; in the month after Maria, only $31.6 million was raised.[342] The US Chamber of Commerce Foundation Corporate Citizenship Center tracks the number of companies that donate to disaster relief. Around three hundred companies donated to Texas during Harvey, compared to only thirty-eight companies post-Maria.[343] Media coverage can also help drive volunteerism during the response to a disaster and afterward in recovery as people learn of the disaster, the need for help, and where to go to volunteer through reporting.[344]

Help always converges from the outside during disasters and catastrophes, but in this case, all of Puerto Rico was impacted. No neighboring towns were unscathed. Even if someone had been able to help, moving freely across the island was difficult because of fuel shortages, blocked roadways, and damaged infrastructure. For those outside Puerto Rico who wanted to help, there were significant logistical challenges. The delays I had faced going to Texas during the Tax Day Flood were nothing compared to the challenges of getting into Puerto Rico immediately following the storm. It took two days for the airport in San Juan to reopen, and even then just to military traffic.[345] It took three days after the storm for the main port in San Juan to accept its first shipment of aid.[346]

As President Trump astutely noted, Puerto Rico "is an island surrounded by water, big water, ocean water."[347] This meant that during Maria, the convergence of help that we are so used to seeing did not unfold in the way it did during other storms like Harvey, Irma, and Sandy. In Texas, when calls were put out for private boats to come help with rescues, people could drive in from other parts of the state and Louisiana. When Irma knocked out power to 4.4 million customers in Florida,[348] thousands of utility workers and their equipment were already standing by in Georgia ready to start repairs as soon as the storm cleared. In Puerto Rico, an ocean stood between survivors and the outside help that was desperately needed.

Because people were not able to physically get there, Puerto Rico was left without the benefit of immediate regional support. What the government and formal response agencies do during a catastrophe is important, but the unplanned help that arrives is often just as much, if not more, of a contribution to the response. In Puerto Rico, this huge piece of the usual emergency management system was missing.

The federal government dragged their heels in the days after the storm on everything from waiving the Jones Act to deploying USNS *Comfort* and other federal resources that could have met Puerto Ricans' life-saving needs.[349] Altogether this wasn't surprising. A government that does not function in nondisaster times will not suddenly become effective when catastrophe strikes. The federal government had undergone a tumultuous year by many measures. Federal agencies were slow to have leaders appointed, a clear vision evaded the administration, controversy and resistance followed every decision, and various forms of distrust had deepened among the government, media, and the public.

On the president's one and only visit to Puerto Rico thirteen days after the storm, Trump claimed only 16 people had died, saying, "if you look at a real catastrophe like Katrina and you look at the tremendous hundreds and hundreds and hundreds of people that died, and you look at what happened here... You can be very proud."[350] In the following days the official death toll rose to 94. Six months after Maria, a Harvard Study estimated the death toll could be upward of 4,645 people.[351]

Throughout his presidency, Trump continued to deny the death toll in Puerto Rico, Tweeting, "3,000 people did not die in the two hurricanes that hit Puerto Rico... This was done by the Democrats in order to make me look as bad as possible."[352] On the contrary, anyone who had been paying any attention knew ninety-four deaths was a gross underestimation, which is why researchers began investigating the real number and demonstrations like laying pairs of shoes in front of the Capitol to draw attention to the undercounting emerged in the months after the storm.[353] The Harvard Study lent credibility to activists, survivors, and disaster experts who had been fighting for a more accurate count to be conducted.

It is not uncommon for post-disaster death tolls to be incorrect.[354] Tallying the number of dead is not as straightforward as it sounds. There can be confusion over what qualifies as a disaster death, as the determination is largely left to the discretion of local coroners.[355] There can also be nefarious attempts at either underestimating or exaggerating the final count, something we experienced on a grand scale during the COVID-19 pandemic. The exact number killed from both direct and indirect impacts of Maria will likely never be known, but regardless, the estimated total is more than should be possible from a hurricane.

The fight over the death toll was just the tip of the iceberg in Maria's response failures. During most disasters, the only operational action a president must take is signing a Presidential Disaster Declaration. Regardless of the president's operational role, there is an expectation that they behave compassionately, in the most general sense of the word. This is even more important in a catastrophic event when leadership from outside the impacted area is needed.

It's not unreasonable to expect this kind of leadership from the president.[356] At the very least, most citizens expect a president to support rather than criticize disaster survivors. Instead, Trump used Hurricane Maria as an opportunity to disparage Puerto Ricans. Just days after Maria made landfall, he Tweeted about the importance of dealing with Puerto Rico's "massive debt" and shifting the narrative of blame solely to the island's weak infrastructure rather than the slow federal response. He repeatedly attacked the mayor of San Juan, Carmen Yulín Cruz, accusing her of "poor leadership" and doing a "poor job" dealing with the storm, and falsely accused the media of disparaging first responders and taking "spirit away from our soldiers." When Trump did go to Puerto Rico to meet with local leaders and receive an update on the response he, inexplicably, threw rolls of paper towels into the crowd.

From the start of his administration many were concerned

about how FEMA, under the chaotic Trump administration, would manage a major disaster or catastrophe. While much of emergency management operates outside the White House, decisions and priorities across the federal government and Congress have direct implications for the system. Overall emergency management made it through Harvey and Irma relatively well. Certainly, some decisions deserve to be critiqued; a perfect disaster response is not possible, but in *general* these disasters were handled in a similar way as they would have under other administrations. When Hurricane Maria struck, the deadly implications of chaotic federal leadership became clear.

In the weeks and months that followed Maria, parts of the emergency management system seemed to all but collapse. Puerto Ricans held a march in Washington, DC, in November, 2017, to demand financial support, one obvious indication the system was failing.[357] The international community even began to intervene, reminiscent of Hurricane Katrina a decade earlier. Oxfam International, which rarely responds to US disasters, issued a statement saying they were "outraged at the slow and inadequate response the US Government has mounted."[358] Even with a functioning federal government and a president who was interested in disaster relief, the situation in Puerto Rico would have been dire. Even on its best day, FEMA is not prepared to respond to catastrophic events.

As the response unfolded, FEMA administrator Brock Long made the media rounds, assuring the American public that FEMA was working nonstop in Puerto Rico. Graphics were shared on FEMA's social media feeds in an attempt to showcase their efforts. In one early update, posted in October of 2017, FEMA described the work they had been doing on the island. They noted the amount of debris their contractors had cleared could fill Yankee Stadium seven times; they had done more than seven hundred airdrops; and they had responded with over 10 million liters of bottled water. The graphic also claimed that the fourth largest cargo plane in the world had transported generators to the island. Forty-two thousand tarps had been distributed to cover roofs and prevent further damage. One hundred thirty-four generators had been installed. The fine print insisted this was more than what was installed during Harvey and Irma combined. These numbers on their own sound impressive, but when placed in the broader context of need in Puerto Rico, they did not make a dent.

The graphic did not make clear if these bottles of water had actually been handed out, delivered to the island, just ordered, or were sitting on an empty runway, as was later discovered to be the case. It didn't make clear what had been air-dropped, where, or to whom. It didn't say how many homes still needed tarps. It did not say, despite the world's fourth largest cargo plane chauffeuring generators, how many people were still without power, or that there had been nearly no need for generators during Harvey and Irma. Of course, Puerto Rico was sent more. Texans had rarely lost power and many Floridians already owned their own generators. An analysis that compared the US federal response to Harvey and Irma to Maria confirmed that these efforts in Puerto Rico paled in comparison.[359] And another report after Maria found FEMA lost 38 percent of the commodity shipments to Puerto Rico, those that did arrive took an average of sixty-nine days to reach their destination, and it took ten days for FEMA to get the first deliveries of food and water to residents in Puerto Rico.[360]

Sure, FEMA was working to address needs in Puerto Rico, but it wasn't being done efficiently or effectively as everyone including the governor of Puerto Rico, the mayor of San Juan, Chef José Andrés, celebrities like Lin Manuel Miranda, volunteers and nonprofits, a national media that was finally beginning to pay attention to the severity of the situation, and thousands of Puerto Ricans pointed out.

FEMA was assessing their effectiveness based on how much

help they were giving, while everyone else was measuring FEMA's effectiveness in terms of how many people still needed help. It's nice that enough debris was removed to fill Yankee Stadium seven times, but that didn't address the problem that many roads were still not cleared. The expectations of many in the public were that the government was meeting all of the needs of Puerto Ricans, but FEMA, and the formal emergency management system more broadly, was not even attempting to do so.

FEMA Administrator Brock Long was right that they were helping, but everyone else was also right that the federal government wasn't doing enough. The government wasn't meeting everyone's needs, and even if they had wanted to do more, which some at FEMA certainly did, they weren't built to do more. I suspected some of those "FEMA: Fix Everything My Ass" shirts from New Orleans would be showing up in Puerto Rico soon.

I had flashbacks to Hurricane Katrina and the levee failure as I watched the slow trickle of aid finally start to flow into Puerto Rico. I saw many who were confused about why the federal government, especially FEMA, was failing to provide adequate help when they just had done so in Texas and Florida. As destructive as Harvey was and as big of a response as Irma required, both fell comfortably within the existing response plans and systems. Maria did not. As they had during Katrina, the federal government showed up in Puerto Rico with their disaster plans, not their catastrophe plans.

The events of 2017 pushed the US emergency management system to the brink. One place after another was overwhelmed by crisis, and the broader system strained under the widespread need. The US had to contend with Harvey, Irma, and Maria in the span of just a few weeks. In the midst of the response to Maria, fifty-eight people were killed and hundreds wounded in the Las Vegas shooting, which earned the title of the deadliest massacre in modern US history.[361] The end of the year also saw multiple wildfires on the West Coast, including the Tubbs Fire, at the time the largest in California history.[362] In December an Amtrak train derailed in Washington State. The car dangled from an overpass onto the highway below. Three people were killed and the incident reinvigorated discussions about transportation safety. On our very best day, and with a competent administration, so many crises in such a short period of time would present major challenges.

Fast and effective response to minimize loss is what our emergency management system is supposed to be good at doing. But by the end of 2017 that system was already struggling to keep pace. By the end of October, FEMA was spending $200 million dollars a day on response and recovery efforts for the three record-breaking hurricanes and deadly wildfires in California.[363] At year's end, initial tallies of disaster damage nationwide topped $320 billion with around twenty disasters topping $1 billion.[364] In November, Brock Long testified before Congress that the agency was "tapped out."[365]

Following the 2017 hurricane season, researchers found that FEMA did not have the resources or personnel to respond to simultaneous disasters, and that this was a contributing factor in the failed response to Maria. One poignant example was an investigation into the "meals" FEMA had provided after the storm. Puerto Ricans reported the "meals" they received were really just snacks made up of assorted junk food. A report from the Department of Homeland Security's Office of the Inspector General confirmed this was what had been provided and specifically found that the reason FEMA was unable to supply sufficient and nutritional meals was because their contracted vendors were already overwhelmed meeting the needs in Texas from Harvey.[366]

There is no such thing as a perfect response to a catastrophe. The very nature of catastrophe means needs will go unmet, help will be slow to arrive, and coordination will be challenging, if not impossible. When we look at the factors that created Maria, it is clear that there was plenty that could have been done long before the storm ever formed to save lives and lessen destruction. More could have been done to decrease Puerto Rico's vulnerability long before there ever was a hurricane. There could have been better plans and capacity could have been developed for simultaneous deployments. There could have been federal leadership that supported local leadership. There could have been better planning in the years before the storm—it's not like it was a surprise that Puerto Rico is surrounded by water.

A few months after the 2017 Atlantic hurricane season, a reporter asked me what had surprised me about Harvey, Irma, and Maria. It broke my heart to say, "Nothing." The responses and recoveries have played out exactly as I would have expected given what decades of disaster research tells us. "We didn't know this would happen" is not an acceptable excuse. Maybe it is good I was not surprised, because it means that we have the knowledge to change. We know how to manage disasters more effectively, efficiently, and justly—not doing so is a decision that is being made. The challenge now is finding the courage and political will to make the changes we know are needed.

Our risk has increased, and we have not grown our emergency management capacity at a commensurate rate. The United States is not prepared for many disasters and we are especially not prepared for the disasters of the climate change era. We will remain until policy changes are made. If we want to minimize and prevent disasters, we have to make informed development decisions that put the public's safety over profit, mitigate climate change itself, address inequality, fund community mitigation projects like the ones needed in Camp Ellis, and be honest about our risks and capabilities. In doing so we must acknowledge that disasters will happen and ready ourselves to manage them when they do. We have to do more and be better.

Maria made visible the weaknesses in US emergency management. The decades of work building one of the world's most advanced emergency management systems, the changes that the federal government swore had been made post-Katrina, the so-called lessons they promised they had learned, were revealed to be exaggerations of the tallest order. Adding to the injustice was the failure of the national media to quickly, fully, or accurately cover the catastrophe. The people who survived, and the many people who did not, never received the attention they deserved and needed. Maria was the second catastrophic US hurricane of the twenty-first century, but it was also a critical warning that the Trump administration was not capable of leading a response to a crisis. The strain in 2017 was a warning that our emergency management system was not ready to face future crises. It signaled that we were not prepared for catastrophic events or multiple disasters occurring simultaneously—as we all got to experience together in 2020.

PART FIVE

Preparedness: Anticipating Disaster

*"I don't know about you, but I intend to write
a strongly worded letter to the White Star Line about
all of this."*

—LEONARDO DICAPRIO, *TITANIC*

IN THE FALL of 2019 I moved to Omaha, Nebraska, for a new job. I didn't think it would be possible to survive another Fargo winter, so although living in Nebraska had never been on my bucket list, it was the warmer option. Before I moved, I visited Omaha to find a place to live. I looked at one bad apartment after another until I finally stumbled upon one, at the last minute, that had a sweeping view of downtown Omaha. I moved in and watched the sunset every night over this little city in the dead center of the country.

From my apartment, I looked down on a park 'revitalization' project and the *Omaha World-Herald* building that kept their lights on well into the night. I could see the Old Market and off in the very distance, on a clear day, I could almost see the edge of the University of Nebraska Medical Center (UNMC). When I moved, my attention was mostly focused on the flooding rivers. I had no idea that just months later Omaha would be the center of another crisis—the COVID pandemic.

★ ★ ★

In early January my disasterologist senses began tingling. I had been keeping half an eye on the unfolding situation in Wuhan, China, and saw enough to know COVID-19 was something to monitor. By the end of the month, China had built a new hospital from the ground up, a sure sign of how serious COVID was.[367] It is easy to succumb to the belief that "it can't happen here" even when there is little evidence that is true. Almost everyone I mentioned it to in January waved it off saying it would just be a problem for China. In a way, I believed this too.

It was not that I thought COVID would not come to the US—that was all but a guarantee given air travel. But, in January I wasn't too worried because I thought we could manage it—we had a heads-up about what was coming. One of my first outings in Omaha had been to a "Science Café," a monthly event hosted by UNMC that gives researchers from various disciplines the opportunity to share their work in the casual setting of a local bar. The one I happened to go to was on the Nebraska biocontainment unit at UNMC that had famously been used to treat Ebola patients in 2014.[368]

I sipped on a cider while the clinical coordinator explained the history and capabilities of the unit. It was just a small glimpse into the work that they had done in the past few years, but it was impressive. On my way home that night, though, I worried about scale. Their unit was effective for treating a few patients, but it was clear there was little ability to scale up their operations to treat hundreds, thousands, or more. Still, there was something comforting about knowing the leading experts in the country—if not the world—were just a few blocks away.

This is why, throughout January and the first weeks of February, when I looked out my apartment window over the city of Omaha, Nebraska, I thought that although there was a budding health crisis, it was under control. In early February, fifty-seven Americans were evacuated from Wuhan and brought to Camp

Ashland, just outside Omaha, to be quarantined.[369] The night they landed, the city filled with sirens as they were escorted from the airport. A week later, several Americans from the Diamond Princess cruise ship were brought to the National Quarantine Unit and the Nebraska biocontainment unit.[370] The investment that had been made to transform UNMC into a global leader in bio-preparedness seemed to be paying off. I remained vigilant and concerned about the possibility of more widespread impacts in the United States, but when I looked out my window (quite literally), I saw the situation being taken seriously and managed appropriately.

From my perspective, I saw signs that the public health systems that had been put in place for exactly this scenario were working. There is a difference between an ongoing crisis that is being managed and an ongoing crisis that is not being managed. What I saw made me think we were in the former. Sometimes it is only in retrospect that you can understand the full extent of the mismanagement.

The National Response Framework designates the Department of Health and Human Services (HHS) as the lead agency for public health emergencies.[371] This approach is largely replicated at the state and local level, with emergency management agencies expected to serve more of a supporting role to public health agencies. Specifically, within HHS, the Office of the Assistant Secretary for Preparedness and Response (ASPR) was supposed to be the lead in providing expertise for this type of situation. The way our plans were written made me feel that the prevention and response efforts for COVID-19 fell much more within the purview of public health, rather than emergency management. So when the first US case of the virus was announced on January 20 in Washington State,[372] it still seemed that the situation was a public health problem, not an emergency management problem.

By the end of January, cases were increasing and the World Health Organization declared a public health emergency. I felt the shift that comes from such a declaration. The group texts among my emergency management and public health colleagues began to light up as we forwarded email alerts and news articles to one another. I could feel our collective disaster vision start to kick in as we started to run down the various scenarios—some that played COVID out as a short-term minor inconvenience, while others were so catastrophic that I won't write them here.

Soon my emergency management students began asking me questions about COVID during class. I told them what my public health colleagues told me; this was a serious situation, especially in China where case numbers were growing quickly. It was something that we needed to watch closely in the United States, but as they had been learning from me all semester, even though it is not perfect, we do have extensive infrastructure to manage disease outbreaks and other crises. In understanding the wide range of possible scenarios, I landed on a noncommittal line that I repeated again and again: this is not the apocalypse, but it is serious.

Midday on March 1 my phone buzzed with a push notification from the *Washington Post*. I opened it to see a headline that took me a few seconds to fully digest and come to terms with. The headline read: "Coronavirus probably spreading for six weeks in Washington state, study says." This changed everything. That familiar tug returned to my chest and all I could think was, "fuck."

I realized that although the known cases of COVID may have been managed well so far in the United States, we did not know how many cases there were. This new information allowed me to knock out the less threatening COVID scenarios that I had still been hoping were a possibility. From the outside looking in, there was plenty about the government's response—or lack of response—that the public didn't know about or understand at the time.

For instance, what had not yet been made public was that scientists in Washington State suspected there was ongoing community spread but were unable to test for COVID. Dr. Helen Y. Chu, an infectious disease expert who had been gathering samples from patients in the Puget Sound area since January for a study on the flu, realized the samples could be used to test for COVID-19. She asked state and federal officials for permission to repurpose the samples but was told it would not be allowed. On February 25 Dr. Chu and her team ran the tests anyway, without their approval. The daring decision revealed a previously unidentified case of COVID-19, which suggested that the virus had been spreading in the United States for weeks.[373] It would take even longer for the full picture of testing failures to come to light, but the end result was that the United States had not been testing as COVID spread across the country. The call was coming from inside the house.

The *Washington Post* headline also signaled to me that I had fallen into a classic disaster trap—I thought because I did not see the disaster, that there was no disaster. The morning after Hurricane Irma passed through the Caribbean as a Category 5, damage reports made their way into the national news.[374] Conspicuously absent was any word from the island of Barbuda, home to 1,600 people. There were no posts to social media, the Snapchat map was empty, and phone calls were not going through. Even the ham radio folks were unable to reach anyone. Many said communications were just temporarily down and everything was probably fine. When Prime Minister Gaston Browne finally flew to Barbuda he found near complete destruction. Weeks later he addressed the UN General Assembly and explained, "The footprints of an entire civilization have been emasculated by the brutality and magnitude of Irma."[375]

In the first few hours after major disasters, before photos

of the damage make their way to newspapers' front pages and Twitter, those on the outside may not even realize a disaster has happened. In the absence of photos or reporting of a crisis, the default assumption is that there is no crisis. But often, what the lack of information indicates is that people are not posting photos because power lines and cell service are down, and they have no way to tell the outside world of the disaster. Not that everything is fine. In a disaster, silence is the scariest sound.

This is what happened with COVID-19 in the United States. There was an assumption among many in the public, early on, that a lack of news was a sign that things were fine, when in reality it should have signaled an invisible crisis was already underway. The lack of testing obscured what was actually happening. It wasn't until there was confirmation that there had been community spread for a significant amount of time that the silence was broken.

An alternate timeline of 2020 was possible early on; the best-case scenario that my disaster vision allowed me to see. It is a version where we had widespread testing, tracing, and isolating beginning in January. It is one where people heard and acted on the alarm bells rung by public health experts. It is one where US leadership, alongside WHO, prevented the crisis from becoming a pandemic. A timeline where US deaths did not reach into six digits and we did not go through this turmoil, fear, and pain. It is a timeline where the White House, HHS, FEMA, and Congress used existing plans to facilitate a responsible and effective response. It is a timeline where the president of the United States didn't publicly downplay the risk of mass death even while telling Bob Woodward it was "deadly stuff." It is a timeline where COVID was an effectively managed crisis, and not the cruel catastrophe that has unfolded.

Surviving on Checklists & Go-Bags

THE RED CROSS, FEMA, and local emergency management agencies all like to put out pamphlets with checklists for how to prepare for a disaster. The crown jewels of preparedness are the "go-bag" and "preparedness kit." The idea is to have gathered all the supplies you will need in the event of a disaster in one place. For some, that means a kit that stays in your house, and for others it's a travel bag that you can quickly grab on your way out the door. During National Preparedness Month, emergency management agencies lead campaigns to remind people to update their kits. Standard items include bottled water, nonperishable food, a battery-powered radio, flashlights and extra batteries, a whistle, wrenches to shut off utilities, a manual can opener, medication, a "deluxe family first aid kit," a multipurpose tool, emergency contact information, copies of important documents, and cell phone chargers.

Emergency managers and disaster nerds *love* to tell you what is in their kits. They always seem to know about the latest gadget or item you would not have thought to include but does seem like it would be really useful. They espouse long-established philosophies of the type of bag they use to store these items and rationales behind every item they've included. The value of these kits and go-bags is on their face obvious, which is why it might surprise you that I don't have one.

Yes, you read that right. I'm a disasterologist without a preparedness kit or a go-bag. I've never publicly admitted this before. On the few occasions I've been asked in an interview, I've dodged the question by pivoting to what someone might want to include in their kit or mention items I've packed when needing to evacuate or brought with me when I've gone to a disaster. I am very aware that the emergency management folks reading this are about to get very mad at me, so let me explain myself.

I have never sat down to intentionally make one of these kits, at least not in the traditional sense. Growing up, my parents didn't have a preparedness kit or go-bags. It was just never something I was exposed to or even knew people had until I started doing disaster work. Once I found out preparedness kits were something that people should have, I always seemed to find some kind of justification for not putting one together.

In New Orleans, I lived in a tiny dorm room and, to be honest, was not willing to concede any of my limited closet space for a bag full of canned food. Supplies for just one person may not take up too much space, but a family of six would need eighteen gallons of water to meet the requirements for the recommended three-day supply. If you have a house with a basement this may be manageable, but for a family who lives in a small apartment this can be significant challenge. By the time I moved to Fargo, I had more space, but I did not have the money for a kit. As a graduate student, I struggled to afford rent each month. A $30 weather radio didn't even register on the list of things I should spend money on. People who are unemployed, underemployed, live paycheck to paycheck, and rely on programs like the Supplemental Nutrition Assistance Program (SNAP) and local food pantries to get by on a regular day cannot be expected to stock-pile extra food and supplies for some hypothetical future disaster. After I graduated, I found that I had both the space and money to have made a kit and a go-bag but I still didn't do it. What I have instead is a purse.

I have yet to be convinced there is much of a difference be-tween the bag I carry around with me every day and a disaster kit. Lisa Miller wrote an article in *The Cut* about how, gener-ally, men walk out the door every morning with nothing but a wallet and a phone[376] while women tend to lug around a purse, if not multiple bags, full of everything we could ever need. It's a relatable article that succinctly argued a point I had been mak-ing for years—I do not need to keep a special preparedness kit in a closet, because I carry one with me every day.

In my purse I always have a Nalgene water bottle, and because I have weird food allergies, I am particularly diligent about car-rying snacks. To be fair, it is not the recommended three gal-lons of water or three days' worth of meals, but I could easily survive off it for a few days in a pinch. I have a flashlight in my purse, not because I want to be ready for a power outage, but because it is built into my phone. In fact, at least 81 percent of US adults have a flashlight on them because that is the number of Americans who have smartphones.[377] I keep a phone charger at home, in my car, in my office, and have at least one external battery in my purse at all times. It's not that I expect a disas-ter to happen every time I walk out the door, but because, as a millennial, the idea of my phone dying in the middle of the day induces anxiety.

My phone actually checks a lot of the boxes on these prepared-ness lists. My contacts list has all my family and emergency con-tact information (I do also have my immediate family's phone numbers memorized because I was born when we still needed to memorize phone numbers). I have Google Maps (which is ar-guably better than having a paper map, which I wouldn't know how to read). I can access copies of all my important personal documents from my phone (do not tell the cybersecurity folks about this). Admittedly, I do not carry around a weather radio in my purse but really, that is just a stand-in for being able to receive information. I am signed up for just about every emergency alert you can get, and many are sent as a push message to my phone. Yes, losing cell service is always a risk, but it is also not something that happens in every disaster and particularly not in the types of disasters I am prone to, where I live. (By the time you read this I will have bought a weather radio so the meteorologists do not get mad at me.)

I do not have a blanket dedicated specifically for emergencies, but as a person who runs cold I am never without a scarf, which is just a bougie blanket. My wallet is always in my purse and while I do not carry around enough cash to buy a plane ticket, I usually have enough for a couple tanks of gas. The Red Cross also recommends a "deluxe family first aid kit." I'll be honest that I am not exactly sure what a deluxe first aid kit includes, but for what it's worth I do have an entire CVS in my purse. Need a Band-Aid? Ibuprofen? Tissues? DayQuil? Benadryl? I could cut the strap off my purse and make you a tourniquet. I don't have a multipurpose tool, as recommended, but I do carry a knife and a whistle in my purse because you can take the girl out of New Orleans, but you can't take the New Orleans out of the girl.

Although I have never sat down to intentionally pull these items together, I can go down the Red Cross checklist and mark off nearly every single box. This is an important lesson about preparedness. Much of what makes us prepared for a disaster is done without us realizing it will make us more prepared for a disaster. I didn't intentionally turn my purse into a preparedness kit. These are just the things I carry out of daily necessity or convenience. It doesn't matter why you have these items, just that you do. In fact, I would argue that keeping them in my purse, which is with me at all times, is actually more effective and efficient than having a separate kit in my apartment, in my car, and in my office, as recommended. I also know I am not alone in accidentally walking about with a preparedness kit 24/7. Have you seen the sizes of purses Target sells?

FEMA conducts an annual National Household Survey (NHS) to assess the extent to which Americans are engaging in preparedness efforts. The 2020 NHS found that 81 percent reported having at least three days of the kind of supplies on these lists.[378] This is great but it also means that somewhere around fifty million Americans don't have these things. The prevalence of preparedness kits and go-bags is a bit of a mixed, well, bag. Some people have basements full of supplies, some have purchased an Eddie Bauer winter weather kit that is sitting in the trunk of their car collecting dust, and some of us are relying on the purse approach.

Furthermore, while the recommended items on these lists are intuitive, they are not actually based on empirical evidence of what makes someone prepared for a disaster.[379] These preparedness checklists originated decades ago and have remained largely unchanged. For example, the theory behind the three days of food and water recommendation was just that it shouldn't take more than seventy-two hours for help to get to you.

More recently some local emergency management agencies have begun recommending people stockpile as much as ten to fourteen days of supplies. It could be the case that a two-week recommendation is appropriate. There is a growing possibility of even longer response times as disasters increase in frequency and capacity weakens, but again, this recommendation isn't based on empirical evidence. Further, the same people who cannot afford the three-day supply recommendation will also be unable to afford the recommendation of even more supplies.

We can nitpick these recommendations all day long, but what is important to understand is that preparedness is about readying ourselves to effectively handle response and recovery. Just because someone has checked these boxes does not mean they are prepared for a disaster. Sure, three days of food and water can help in some instances, but there are many more instances where these are not the actual resources that are needed to survive—a difficult lesson I hope we all learned once and for all in 2020.

In early March, before there were stay-at-home orders, I was in Hawaii at an emergency management conference. Between conference sessions, I walked down the street to the beach. I looked around for some shells and took some pictures of the palm trees, but really, I just doom-scrolled Twitter, which was at this point full of evidence of the federal government's incompetence in handling COVID testing. Even in those early days, it was clear that we didn't have the data we needed to be able to make effective decisions.[380] From the beach, I watched Italy's COVID cases skyrocket and saw the early signs that we weren't far behind. In a café at my hotel, a man standing in line behind me said loudly into his phone that all of this was "mass hysteria." I turned around and passive-aggressively offered him some hand sanitizer.

There was something dystopic about sitting in paradise moments before a pandemic. Though perhaps there is something dystopic about being anywhere in the moments before a pandemic. I felt the same dread that comes in the days before a hurricane makes landfall, when most people do not quite know yet the devastation that is coming, but my disaster vision has kicked into high gear. Through a flood of reporting, I watched helplessly as bad decisions were made, or worse, no decisions were made, as the clock ticked down. There was nothing I could do, and I could not bear to sit through another day of hearing about disasters and climate change, so I ditched the last day of the conference. I snuck into a luxury hotel with a rooftop infinity pool and drank a few piña coladas before I headed to the airport.

I agonized during the flight back to Nebraska. A friend in public health texted me advice to wear a mask (this was before the CDC recommended doing so) but I didn't have one. I kept my face buried in my sweatshirt as I walked through the airport, which, in retrospect, probably wasn't very effective. Anytime I walked by a bathroom I went in just to wash my hands and sat on the floor far away from anyone else before boarding the plane. It was still early enough in the pandemic that I was the only one who Cloroxed their seat when we boarded.

Before I left for Hawaii, I had stocked my apartment with two weeks of groceries, bought cleaning supplies, filled my car with gas, and brought everything I would need home from my office on campus. I called my family and told them to stock up, too. I knew what was coming, so we made it to the stores when the shelves were still stocked with hand sanitizer and the toilet paper supply chain problem hadn't yet happened.[381]

I anticipated that I would voluntarily self-isolate in Omaha when I returned, in case I had come into contact with someone who was sick while traveling. I was fortunate that I was able to afford the extra cost of the groceries and had a job that I could do easily from home. By the time I was flying back to Nebraska I was regretting my decision to travel, but I felt ready for two weeks of lockdown. After that, I would reassess.

While I had done an excellent job of readying myself for two weeks in isolation, I had made the amateur mistake of forgetting that, as the oldest of four, it is my birthright to be responsible for my siblings. The day before I left Hawaii, my preparations were thwarted when I got a call from my brother, a college freshman in Chicago, telling me he had just found out they were shutting campus down for the remainder of the semester. At eighteen he was not old enough to rent a car to drive back to our parents' home in Maine, and we were concerned about him flying as his history of asthma put him in a high-risk category. In a stroke of pure luck my layover back from Hawaii was in Chicago. I disembarked at O'Hare, rented a car, picked him up from his dorm, threw his stuff in storage, and we made the six-hour drive back to Omaha, figuring the risk of him getting COVID from me was less than his risk of flying through multiple major airports.

Even despite my efforts before I left for Hawaii, I was unprepared for a pandemic. And now it was too late. It was still only mid-March, and with conflicting messaging from the White House, it was difficult to anticipate what the next few weeks would look like across the country.[382] Stay-at-home orders were beginning to roll out, but the timing and extent of the orders were hard to predict. I spent the drive back to Omaha reassessing our readiness for the pandemic and narrated a to-do list to my brother. Our most immediate problem was that we would need to stop at a grocery store because my brother, a D1 shot-put thrower, was not going to make it two weeks on my supply of chickpea pasta and cauliflower rice.

The logistical challenges grew exponentially from there. Most pressing was that I was scheduled to be leaving my job in Omaha at the end of the semester for a new job in Boston. Moving across the country in the middle of a pandemic was not something even I had on my 2020 bingo card. My lease in Omaha ended in April, and I had yet to even think about apartments in Boston. After only a week in Omaha, I made the decision to move earlier rather than later in the hopes of avoiding a scenario of getting stuck in Omaha without a place to live. It was a difficult choice and one that was made with great deliberation and many Clorox wipes. Instead of putting more people, like movers and a Realtor, at risk in Boston, I put my belongings in storage and decided to go back to our parents' house in Maine. As many people experienced over and over throughout the pandemic, it is difficult to make responsible decisions in the middle of a crisis, when there is a lack of information and mixed messages coming from government.[383] As is the case during every disaster, we were trying to make the best decisions with the information and resources we had.

This series of unexpected events is how I ended up driving halfway across the country with my brother in the middle of a pandemic. We left Omaha at 3:00 p.m. on a Saturday afternoon with clear blue skies and quickly warming spring weather. In normal times, the sidewalks and park near my apartment would have been full of people walking their dogs. The Old Market would have been bustling with tourists and college students drinking beer and watching sports. Instead, it was eerily empty.

Halfway through Iowa I realized I had been clenching my jaw for the last month. Something about those golden waves of soon-to-be-grain that we passed let me breathe in a way living down the street from UNMC's biocontainment unit had not allowed. The roads were quiet, even as we drove back through Chicago. I had packed my apartment as fast as I could, but we were a few days behind everyone else. There were just trucks, essential workers, and a few other stragglers left on the road. Later cell phone data showed that Americans really did alter their behavior and stay home during these first few weeks.[384] Along the drive, anytime I had almost forgotten what was happening— trying to grasp a fleeting moment of normalcy, we would pass a highway sign reading "flatten the curve" and "we're in this together"—at the time still a novelty.

I had been doing the drive from the Midwest to the East Coast for nearly a decade, but this time it was surreal. We passed through unmanned tollbooths and by shutdown rest stops. We interacted with no one on the twenty-one-hour drive and drove through the night, not wanting to stop at a hotel and put others at risk. As the sun set, my brother fell asleep and I let my disaster vision take over as I imagined our now dwindling number of COVID scenarios. The sheer amount of death that was coming was so great I didn't have enough space in my mind to hold all of the people who would be lost.

On our very best day, managing a pandemic would be difficult, but during the Trump administration I knew it would be a response unparalleled in its incompetence. I tried to hold on to a shred of hope that there was still enough of our institutions left to make up for the administration's ineptitude.

Our attention has largely centered around this traditional concept of preparedness. If you buy just the right supplies, in just the right quantities, you'll be okay. It's an individualistic approach that allows some to buy their survival. The pandemic revealed the limitations of those preparedness checklists. I never needed a flashlight, and fortunately my plans of using a purse strap as a tourniquet has not been needed (so far!). Sure, there were supplies I ended up needing, like masks, but there was no single item that enabled me to survive the pandemic.

What has actually kept me safe during the pandemic is the financial ability to buy what I needed. I could afford to stock up on two weeks of food at once. I was able to pay for the unexpected car rental in Chicago and buy an air mattress for my brother. I could afford to move a month earlier than expected with only a few days' notice. I have job security and have been able to work from home. I have health insurance and was never worried about how to pay for a hospital stay if I became sick. My expertise has enabled me to navigate the constant mixed messages from government officials. I know where to find reliable information and have been able to make informed decisions based on that information. I have a network of people to help me when I need it. None of these things fit into a preparedness kit or a purse.

The pandemic gave us all personal experience that just checking off a handful of items on a list does not mean that we are prepared for a disaster. In fact, it can even give us the false impression that we are ready. It is not bad or wrong to have a preparedness kit, and if you have the resources to make one you should. You need to be careful, though, that it isn't all that you do.

Two researchers, Dr. Jessica Jensen and Dr. Emmanuel Nojang, recently published research that sought to better understand what makes someone prepared for a disaster.[385] In their review of the existing empirical research they confirmed that it's not only stockpiling supplies that makes us ready for disaster—preparedness is more than that. It includes the knowledge that we have to navigate the disaster, whether we have taken steps to minimize disaster impacts, our social networks, our connections and access to technologies that provide warnings and communications, and our physical and mental ability and adaptability. All of these things together depict a much more holistic picture of what it means for each of us to be ready for disaster.

The traditional checklist approach is one that aligns with the "pull yourself up by your bootstraps" philosophy of self-sufficiency. The preparedness tradition in the United States since the 1950s has largely taken an individualistic approach, which is more in step with American ideology than it is with what disaster researchers have found to be true of human behavior in disasters. We aren't on our own when disaster strikes. Disasters are social experiences that elicit collective responses. And, because this is true, preparedness is much more complex than just preparing the individual—we have to prepare the community.

Designing an Emergency
Management System

I ALWAYS HEARD stories growing up about how my great-grandfather Lew was a sign maker in the Chicago suburbs. I met my great-grandmother Julia, a total spitfire, only a few times when I was very young, but otherwise never really had the opportunity to get to know them. When my grandfather (their son) passed away many years ago, a stack of his boxes, full of mementos, ended up in my parents' barn, which was where I was standing a few years ago when I found a newspaper clipping that made all the pieces of my life seem to fall into place.

I had recently graduated with my PhD in emergency management and was home in Maine for a few weeks of summer vacation. My mother seemed to think I didn't have enough to do, so one afternoon she sent me to the barn to sort through my grandfather's old stuff. I picked a box at random and started going through old photos, letters, plane tickets, and all the other assorted bits that you don't want to part with but don't exactly know what to do with.

I picked up a newspaper clipping. It was a brief announcement about how my grandfather, stationed in Austria, would be coming home to spend Christmas with his parents in Illinois. It was an interesting bit of history but not something we probably needed to save so I set it down on the trash pile. As it fell, I caught the briefest glimpse of the words *Red Cross*.

I picked it back up realizing the newspaper had been folded in half and I had read only the first paragraph. Below the fold, I read that my great-grandmother worked for the Red Cross during World War II and taught first aid to the regular and auxiliary police departments outside of Chicago. I was surprised that, given my field of study, no one in my family had ever mentioned my great-grandmother's extracurricular activities, but at the same time, who *hadn't* volunteered with the Red Cross, especially during wartime?

I kept reading and came across a sentence I had to read a few times to make sure I wasn't hallucinating: "Mr. Posner was a Co-ordinator of Civilian Defense." I abandoned the barn and confronted my mother in the kitchen who was (understandably) very confused about what a "coordinator of civilian defense" was, until I explained that they were the precursors to emergency managers. In research I've done since, I've come to learn that "sign maker" was a much later career path. In fact, my great-grandfather did public relations work for the state of Illinois and helped with the creation and implementation of local civil defense programs during WWII. In other words, my great-grandfather was among the country's very first emergency managers, and I had had no idea.

There was a time when government, especially the federal government, did not involve themselves in disasters.[386] Over time, their roles and responsibilities in times of crisis have fluctuated with our evolving risk, public pressure, shifting ideology, and politics. Part of this evolution has been the development of an emergency management system.

In 1927, the Mississippi River overflowed its banks, flooding 16.5 million acres and displacing almost one million people.[387] It was the most extensive river flood in US history and, outside of the Red Cross, there was no formal plan in place to coordinate the response across the seven states affected. The civil defense offices that people like my great-grandfather created did not yet exist, nor did the emergency management agencies of today. Governors from multiple states had requested help from President Coolidge as the damage grew too big to ignore. The Red Cross felt confident in their ability to handle the crisis in some Midwest states but expected the need in Arkansas, Mississippi, and Louisiana to overwhelm their resources. Facing this reality, Coolidge appointed Herbert Hoover, then Secretary of Commerce, to represent the federal government in the affected states.

Despite Hoover's presence in the Midwest, the prevailing belief at the time was still that response and recovery from disasters should largely be a philanthropic effort, rather than the responsibility of government.[388] Hoover's approach to leading the federal response was largely to launch a marketing campaign to drive donations and resources to the Red Cross. Hoover would travel from town to town, informing residents that displaced survivors would soon be arriving and would require their hospitality. He framed the flooding as a battle against nature that required a patriotic response akin to WWI. His efforts galvanized support and donations for the Red Cross, which was the organization that remained most central to the response. By the 1927 flood, the Red Cross had been working in the US for several decades and had come to be perceived as a reliable support for disaster survivors. Millions of dollars flowed to the organization as Americans across the country rose to Hoover's challenge. The

Red Cross used the donations to distribute aid, provide meals to survivors, and assist towns in running "camps" for the displaced.

Hoover was incredibly successful at managing the public narrative around the response. Despite the quite limited role of the federal government in aiding those affected by the flooding, he pressured journalists to cover his involvement in the response and hired photographers to capture images of him at the relief camps. He also suppressed coverage from Black journalists and writers like Ida B. Wells and W. E. B. Du Bois, who were telling a different story of how the response was unfolding.[389]

Their reporting showed that the Red Cross, and by extension the federal government (recall that the Red Cross has a Congressional charter to assist during disasters), had not distributed aid to Black survivors equally or with the same urgency as they had to white survivors. Further, there were widespread reports of racial violence and exploitation of Black survivors staying at the camps. At the height of the Jim Crow era, the well-being of Black Americans was not of great concern to politicians like Hoover, or prioritized by organizations like the Red Cross. The discriminatory dispersal of aid by white-led organizations like the Red Cross was the impetus for organizations like the NAACP to create their own system for aid distribution through the network of Black churches across the country.

Despite all of this Hoover's narrative of his success over managing the response to the flooding prevailed. He turned the footage of him working in the Midwest into a video titled *Master of Emergencies* and launched his presidential campaign. The campaign was successful, and Hoover was elected president in 1928, carrying forty states to his Democratic opponent's eight.[390]

Hoover entered the White House just as two of the largest national crises in US history gripped the country—the Dust Bowl and the Great Depression. His response to both further undermined his undeserved moniker "Master of Emergencies." With most of the country desperate for financial help, the federal hands-off approach to disaster relief was no longer viable. It had taken years, but eventually Hoover's veneer as the "Master of Emergencies" started to crumble. With the help of the northern Black vote, he lost reelection to Franklin D. Roosevelt, who won forty-two states to Hoover's six.

When FDR took office, the public's attitude and expectations about government assistance had shifted, laying the groundwork for the federal government's increasing involvement in domestic disasters.[391] Though in some ways a radical shift from Coolidge's and Hoover's approaches, federal involvement was still limited, especially considering the extent of suffering during the height of the Dust Bowl and Great Depression. Nonetheless, it set a precedent that federal involvement in major disasters was necessary.

Around the time that the federal government began funding the first major wave of social research on disasters and sociologists started doing quick response research across the country, Congress passed the Civil Defense Act, creating the Federal Civil Defense Administration.[392] Cities and states began to open civil defense offices in an effort to ready the public for a possible enemy attack on US soil, which formalized the efforts volunteers like my great-grandfather had undertaken. The Civil Defense Act also called for the creation of plans and procedures, and facilitated federal guidance for efforts like building fallout shelters, the Bert the Turtle "duck and cover" campaign, and stockpiling supplies.[393] Some localities found it necessary to hire a civil defense director, usually a retired service member, to oversee these projects. This connection led to a tendency toward a "command and control" approach to emergency management—not to mention a field dominated by white men with military backgrounds.[394]

As the nuclear threat in the US subsided, local civil defense coordinators found they were kept busy by the various natural hazards that affected their communities. They worked to figure out how they could rework their civil defense plans to manage these other disasters. By the 1970s, civil defense offices were transforming into emergency management agencies and civil defense coordinators into the emergency managers we have today. These new emergency management agencies redefined their missions to focus on all hazards—natural, technological, and otherwise.

At the same time a series of preparedness programs were developed within different agencies across the federal government. There was a lack of clarity about what each program was for and how they fit together. Further, when state officials went looking for help there was confusion about which programs they were supposed to work with. Eventually, the National Governors Association requested that there be a single agency at the federal level that they could turn to in times of crisis.[395] The answer was the Federal Emergency Management Agency.

In 1979, FEMA was created with the mission to address all phases of emergency management—mitigation, response, recovery, and preparedness.[396] The new agency was given limited resources and struggled at the beginning. Changes to federal policy also meant that the responsibility for emergency management now fell to the White House, rather than Congress. This meant that as each new president came into office, with new goals and priorities, FEMA had to adjust. The agency felt the sway of each new agenda as interests in various hazards, visions, and governing approaches shifted from one administration to the next.[397]

When Clinton took office in 1993 he had just watched the first Bush administration face criticism for FEMA's slow response to Hurricane Andrew in Florida. Not wanting to repeat those mistakes, Clinton nominated James Lee Witt, someone with emergency expertise who Clinton knew from Arkansas, to lead FEMA. In Witt, FEMA finally found some stability, and his tenure at the agency became known as the "Golden Years" of emergency management.[398]

Witt revamped the culture of the agency and shifted its strategy to prioritize mitigation. One program in particular, Project Impact, garnered widespread popularity. Its purpose was to fund mitigation projects identified by local communities across the country. The program was beloved by emergency managers and politicians alike, and communities reaped the benefits. One official in Miami-Dade credited the mitigation projects funded by Project Impact as preventing extensive damage following the 2004 and 2005 hurricane seasons, saying that none of the buildings the program had paid to modify "had any more damage than scuffed paint."[399] Despite the widespread popularity and effectiveness of the program, it was short-lived. As the Clinton administration ended so too did FEMA's "Golden Years" under Witt. Project Impact was one of the first casualties of what would become an emergency management bloodbath led by the second Bush administration.

Concerns about specific types of hazards, like the nuclear threat of the civil defense era, have often dictated the priorities of federal funding and policy. In the two decades prior to 9/11, the groundwork was laid for a shift of attention toward terrorism. Bombings in Beirut in 1983 left almost 300 Americans dead. In 1988, 259 people were killed when Pan American Airlines Flight 103 was blown up over Lockerbie, Scotland. In the 1990s, these attacks made their way to the homeland. A bomb beneath the World Trade Center exploded in 1993, killing 6 and injuring hundreds. Two years later, a similar tactic was used in an attack at the Alfred P. Murrah Federal Building in Oklahoma City, killing 168 people.

On January 31, 2001, the Hart-Rudman Commission released a new strategy for US national security in the twenty-first century.[400] The commission identified the probability of these types of terrorist attacks as an increasing threat to national security, and suggested policies for addressing their growing risk.

Their primary recommendation was to combine twenty-one existing federal departments and agencies into one, tentatively titled the National Homeland Security Agency. Representative Mac Thornberry (R-TX) proposed a bill that would have created this new agency, but found little support among his colleagues in Congress.

Just a few months later, two planes crashed into the Twin Towers and the window was opened for the largest reorganization of the federal government since the National Security Act of 1947. US emergency management was turned upside down, as reckless policy decisions that contradicted decades of disaster research and emergency management expertise were implemented.[401] The cataclysmic post-9/11 policy changes laid the groundwork for the failed response to Hurricane Katrina and the levee failure only a few years later,[402] and still contribute to slow and ineffective responses today.

The most significant of these changes was the creation of the Department of Homeland Security. A month after 9/11, Senator Joseph Lieberman (D-CT) and Senator Arlen Specter (R-PA) cosponsored a bill that took inspiration from the concept of the National Homeland Security Agency, proposing the creation of the Department of Homeland Security (DHS). As had happened when Representative Thornberry introduced nearly the same bill in the House months before, hearings were held, but no vote taken. The Bush administration, with support from the Hart-Rudman findings, presented a similar bill to Congress in June 2002. With pressure from the White House, the bill finally passed, and DHS was created. FEMA was subsumed by DHS along with other agencies, including the Coast Guard and Secret Service. In the reorganization, FEMA lost its status as an independent, cabinet-level agency. Across the country, emergency management agencies got whiplash.[403]

As this drastic reorganization and refocus of the federal government was underway, Congress wanted answers about the circumstances that led to the 9/11 attacks including an analysis of how intelligence agencies had failed to communicate with one another before the attack. Most relevant for emergency management, Congress also wanted a better understanding of the effectiveness of the response at the Pentagon and in New York City. The answers to these questions were compiled into the 9/11 Commission Report released in 2004.[404] The 9/11 Commission's findings concerned many emergency management experts.[405] The Commission found that the immediate life-saving response measures conducted at the Pentagon had been successful. On the contrary, they criticized the response measures in New York City as uncoordinated.

This conclusion ran counter to the findings of disaster researchers in important ways. Disaster scholars documented first responders running into the towers, despite knowing they might never run back out. Inside the towers, people banded together to help each other out of the buildings, carrying coworkers who were disabled down dozens of flights of stairs, resulting in upward of sixteen thousand people evacuating in just 102 minutes. Those who died were largely located on the floors above where the planes collided, making evacuation impossible.

Those who escaped the towers found people outside ready to guide them out of the dust. Local businesses opened their doors to offer refuge, and others handed out water or bandaged wounds. As every bridge, tunnel, and train in and out of the city was shut down, every boat in the Hudson descended on Manhattan. Their efforts resulted in the spontaneous evacuation of half a million people in just over nine hours, making it the largest waterborne evacuation since Dunkirk in World War II.[406]

Despite the loss of the city's state-of-the-art emergency operations center (EOC), located next to one of the towers, New York City's emergency management agency did not falter. Dr. Kathleen Tierney wrote, "both the management and the conduct of emergency response activities continued uninterrupted" and that the "response organizations in New York City were highly resilient, showing great capacity to mobilize and coordinate resources."[407] Formal response operations, many of which were improvised, were run from alternative locations. As outside agencies arrived, they were largely integrated into the already established work being done.

Existing plans and procedures were not perfect, and communication did break down at various points, but considering the scope of the situation, the response itself was remarkably effective. Nearly everyone who physically could have escaped the towers got out. First responders did not abandon their jobs— instead, many sacrificed themselves to save others. The city did not descend into panic or chaos. New Yorkers made spontaneous but rational decisions to help themselves and one another. Given the physical limitations, search and rescue went well, and the immediate needs of survivors were largely met. By these measures, disaster researchers argued the response itself was adaptive and effective.[408]

Government officials, though, measured response effectiveness differently. They were focused on how well coordinated formal agencies were and how well they abided by existing plans. As a result, the need for and presence of a spontaneous response was, to government officials, indicative of failure. They argued a more centralized approach to emergency management was needed. It just so happened that those changes nicely aligned with priorities of the Republican party's long-desired and newly created Department of Homeland Security.

The Homeland Security Act of 2002 decimated FEMA's authority.[409] It was demoted to one agency among many within DHS, and in the process lost the relationships that had been built with other leaders across the federal government, which facilitated interagency coordination and communication in times of crisis. They also no longer had a direct line to the president, instead having to go through the DHS secretary to approve the use of federal resources during ongoing disasters. FEMA's position within DHS also left the agency vulnerable to decisions made by the DHS secretary.

Additionally, FEMA was required to lead nationwide emergency management policy changes that required every single response agency and organization in the country to undergo retraining on a new approach to coordinating response.[410] The changes altered the established emergency roles of dozens of federal agencies, state agencies, local government, and the private sector, and were widely considered to be confusing, as they often failed to explain the exact responsibilities of each agency. Further, many local agencies were particularly resistant to the new mandate because the federal government did not also provide the resources for them to make the required changes. Retraining the entire country on a new, nonintuitive, complex system of coordination was a massive undertaking and did not, at first, go well.

These dramatic changes were implemented in complete disregard of disaster research. Disaster scholars immediately spoke out about the harm these reckless changes would bring.[411] Dr. William Waugh and Dr. Richard Sylves, leading disaster policy experts, warned in 2002, "If the war on terrorism inadvertently undercuts or distorts an emergency system designed to deal with so-called routine disasters, it may well weaken current capabilities to manage conventional hazards and the hazards posed by terrorism."[412] The concerns of researchers were echoed by leading emergency management experts including former FEMA director James Lee Witt, who testified before Congress in 2004, saying, "I assure you that we could not have been as responsive and effective during disasters as we were during my tenure as FEMA director, had there been layers of federal bureaucracy between myself and the White House."[413]

The White House and Congress did not address these concerns, nor did they wait for the findings of either the Commission or disaster researchers before reshaping emergency management. Instead, they had taken advantage of a moment when Americans were scared and Congress would not say no to implement their pre-9/11 homeland security agenda at the expense of emergency management. The changes were based on a regressive understanding of the country's day-to-day risk and have, as the experts warned, made us less safe. Politicians saw a window of opportunity and jumped through it.

It was this beaten, demoted, shell of an emergency management system that collided with centuries of bad policy decisions, a twenty-foot storm surge, and shattered levees in New Orleans. In fact, the government response to Katrina affirmed the very concerns that had been raised by emergency management experts about the post-9/11 changes.[414] Once I learned this history, everything I had seen in New Orleans began to make a lot more sense.

In 2005, the new approach to response was only a little over a year into its implementation and had created dangerous confusion over the responsibility of various agencies, as well as FEMA, DHS, and the White House.[415] Hundreds of agencies descended on New Orleans before they had adequate time to fully train on the new approach. Further, there was miscommunication and confusion between the roles and responsibilities of FEMA administrator Brown, DHS Secretary Chertoff, and President Bush.[416] A Congressionally sponsored analysis on the response to Katrina, "The Federal Response to Hurricane Katrina: Lessons Learned," found that the nation, "at every level—individual, corporate, philanthropic and governmental…failed to meet the challenge that was Katrina."[417]

Following the failed response in New Orleans, Dr. Kathleen Tierney, then director of the Natural Hazards Center at the University of Colorado Boulder, submitted testimony to Congress that explained how the post-9/11 changes had significantly weakened FEMA.[418] The International Association of Emergency Managers agreed, stating that "mixing the DHS mission of preventing future terrorism events and the FEMA mission of disaster consequence management…[had] significantly detracted from both missions."[419] Some politicians seemed to agree. In the days following Katrina, then-senator Hillary Clinton told reporters they would need to "untangle all the causes for the problems" and specifically cited personnel and funding cuts, along with the placement of FEMA within DHS as a concern.[420] By 2009 hearings were held on the FEMA Independence Act and again the consensus that FEMA should be removed from DHS was made clear.[421] Unfortunately, the bill was never voted on.

Congress eventually passed the Post-Katrina Emergency Management Reform Act (PKEMRA), which included three hundred individual changes to federal emergency management.[422] Many of these changes were not as impressive as they first seemed. In the years since, following major disasters, Congress has passed additional bills that have brought some changes to emergency management. Following Superstorm Sandy there was the Sandy Recovery Improvement Act, and after the 2017 hurricane season Congress passed the Disaster Recovery Reform Act of 2018. These have been far from comprehensive reform. Instead, they have brought about more piecemeal changes that were a reaction to some specific and idiosyncratic "lessons learned" from the most recent disaster.

Despite these small changes, the US emergency management system, especially FEMA, remains dramatically unprepared for many of the challenges faced by the United States. In her post-Katrina testimony, Dr. Tierney gave a stark warning that has, in the past fifteen years, been proven prophetic: "My professional assessment is that the nation's emergency management system has been compromised to a degree that the road back will be very difficult."

Planning for Walruses

ONLY FIVE YEARS after Katrina and the levee failure, Louisiana once again found itself at the center of a crisis for which there was no real plan. In the days following the Deepwater Horizon explosion, government officials and journalists dug through BP's government-approved response plans. What they found were what disaster researchers call "fantasy plans"—those plans that may look good on paper but have little utility when crisis strikes.[423]

The plans were outdated and insufficient for the situation, with no instructions on how to plug the hole gushing oil a mile below the surface of the ocean.[424] The wildlife expert they included in their response plans had died four years before the plans were approved by federal officials at the Minerals Management Service (MMS), the agency responsible for industry oversight. The phone numbers of other experts who were expected to help were wrong. Even the calculations used to estimate spill volume were incorrect. Most ridiculously, the plans, *for the Gulf of Mexico*, included instructions for how to remove oil from walruses.

The walruses were the dead giveaway—BP had just copied and pasted their emergency response plans.

In the absence of useful plans, officials started looking around for help. BP solicited suggestions for how to plug the hole on their website[425] and James Cameron (as in the director of *Titanic*) held a brainstorming session in Washington with leading scientists and engineers in an effort to figure out a way to stop the leak.[426] Even Donald Trump, then just the host of *The Apprentice*, called the White House Chief of Staff, David Axelrod, and offered to lead the response.[427] President Obama appointed Coast Guard Admiral Thad Allen instead.[428]

While much of the response to the BP disaster was improvised, the planning and preparedness efforts that government had been developing for years were used to an extent.[429] The response utilized the National Incident Management System, the Incident Command System, and other tools of emergency management. Representatives from multiple federal agencies, the president, state and local officials, and BP officials were all involved in the response efforts. The response as a whole was far from perfect, but the underlying hallmarks of effective emergency management were present.

The traditional preparedness recommendations given to individuals like keeping a few days of food and water on hand were not particularly helpful in this scenario. The people affected by this disaster were reliant on BP to have created effective emergency plans and for government to hold them accountable. Nothing about the BP spill, however—from the monitoring of oil, cleaning up oil, organizing for political efforts, health impacts, economic impacts, and lawsuits—was an individualistic effort.

Although preparedness has historically been framed as an individual responsibility, the reality is that the actions taken by government and the private sector are central to our ability to prepare for disaster. When the response to a disaster doesn't go well, it's not because you don't have enough bottled water stored away in your coat closet. It is because our policies do not prioritize risk reduction and our emergency management system has perpetually been a political bargaining chip that, even at its height, has never received the investment it needs.

One of the only pop culture depictions of a local emergency manager is Tommy Lee Jones in *Volcano* (1997). His character is portrayed as a one-man, lava-diverting hero who has the primary decision-making authority for the county of Los Angeles. His character represents what most emergency managers probably wish they did, but it is far from reality. It is actually Don Cheadle's character, the deputy emergency manager, that is a much more accurate, albeit less exciting, depiction of the job. While Jones is climbing through downtown LA sewers, Cheadle is in the city's emergency operations center fielding calls, directing resources from various agencies, and filling out paperwork. Although it's not the world's most glamourous job, emergency management officials play a critical role in disaster response. They also play a critical role in preparedness and mitigation long before disaster strikes and in the recovery afterward. This isn't an easy job and it can be made more difficult when emergency managers aren't given the resources they need.

The problems that permeate FEMA at the federal level are often replicated at the state and local levels. Despite knowing that investment in mitigation and preparedness pays off, emergency management agencies across the country are perpetually underfunded, particularly in rural and poor communities that are unable to dedicate the resources to the work that most know need to be done.[430]

Local emergency managers often have very little power. Legally, the authority for decision making tends to fall to the mayor or other elected officials. Further, rather than being stand-alone agencies, many local emergency management offices are located within police and fire departments where they have to navigate a hierarchy of people in order to even talk to decision makers.[431] Even more problematic is the fact that their position within these departments leads to an emphasis on response, rather than comprehensive emergency management that factors in mitigation[432] and recovery.[433]

In the summer of 2020 during the height of the protests over the brutal murder of George Floyd by a Minneapolis police officer, Los Angeles mayor Eric Garcetti released the 2020-2021 city budget. There was widespread outrage as the proposal showed $3.14 *billion* earmarked for police. In comparison $6 *million* was earmarked for emergency management.[434] Tommy Lee Jones and Don Cheadle are going to need more than that to save LA.

Some emergency management agencies in big cities are more robust. New York City Emergency Management has over two hundred employees and a $40 million budget,[435] but that is not the reality for most emergency management agencies across the country. Many rural communities and small towns have only a part-time or *volunteer* emergency manager.[436] Smaller communities that do not have the budget or need for big agencies often rely more heavily on county agencies. Given the lack of resources and support, local emergency managers are limited in what they can do to prepare their agencies and their communities for disaster (let alone work on mitigation and recovery).

Planning is one obvious starting point for helping us be more prepared.[437] Yet, even this relatively simple task can be a struggle for some agencies. Response plans, like those produced by BP, are completely inadequate.[438] Although the plan quality varies, most if not all communities have some type of response plan. They might not have been updated recently and may just be sitting on a shelf, but they are something. It's not enough, though, to make a plan to be able to check a box and say you did. When this happens (and it happens more often than we would like to think), it creates huge problems. Planning is a continual process, not something that can just be done once.[439] Having a dusty plan sitting on a shelf does not make anyone safer.

If a plan is to be effective, the planning process cannot be done by an emergency manager sitting alone in their office. It has to include everyone who is a part of the plan[440]—or as FEMA puts it, "the whole community." Research has consistently shown that the most important part of planning is the process, rather than the plan itself. In other words, it is various community stakeholders coming together in the same room to talk about what could go wrong, how they will respond, and who will be responsible. The process itself is what makes planning worth doing, and filling out a "cookie cutter plan" downloaded from the internet—or copy and pasted from your Arctic plans—isn't enough.

It's also not enough to just have response plans. We also need to prepare for *recovery*. Recovery planning rarely happens, despite research that suggests planning for recovery can make the process more effective and efficient.[441] Communities often arrive in recovery exhausted from the response. They end up playing "catch-up" for years, which sets the community up for a difficult recovery process like I saw in New Orleans and other communities around the country. These long recoveries can create divisions in the community, hinder the implementation of mitigation efforts, and leave the community vulnerable to outside interests.

It takes time, expertise, and effort to craft the types of plans that prove to be effective for response and recovery. It requires people who have the time and focus to guide the planning process. Plans also need to be continuously updated as our risks and the people who will be involved in the response are constantly changing. Planning also requires access to the people and technical resources that can provide accurate and useful data.

Even more fundamentally, in order to plan effectively we need to understand our risk.[442] Part of understanding our risk means knowing what hazards we may face, their possible severity, and the likelihood of their occurrence. Emergency managers and public officials often rely on experts to provide this information.[443] Data on earthquakes comes from seismologists, flooding from hydrologists, and terrorism from homeland security experts. Some communities may face legitimate barriers to conducting accurate analyses like a lack of access to data, lack of expertise, and, at times, a lack of imagination.

Assumptions made while compiling risk assessments have to be made on good, accurate, complete data. One frequent issue is that communities base their planning on historical records. In other words, they use the disasters of the past to plan for the disasters of the future. This isn't a good approach as it assumes the worst has already happened. It also falsely assumes that no new hazards have been introduced and that the community's vulnerability has remained static. The exact same hazard occurring in the exact same place will likely produce radically different impacts and needs in the twenty-first century as compared to the nineteenth.

For these reasons, communities cannot ground their understanding of risk solely on the historical record; they must also include our projections for the future, including climate change. Just because your town has never flooded before doesn't mean you won't flood in the future. In addition to climate change, other factors, like development in new areas and changes to the ecosystem (like filling in marshes and bayous), alter the risk a community faces and need to continuously be taken into account.

Since it is not the hazard alone that creates disaster, risk assessments need to consider the built and social vulnerabilities of the community. A community where the business district is located in a floodplain may be particularly economically vulnerable, whereas a city where few people have personal vehicles may face a particular challenge with potential evacuations. Understanding who in our communities is vulnerable, why they are vulnerable, and what their actual needs are is necessary in order to plan effectively.[444] If we do not view a certain hazard as a risk or recognize our vulnerabilities, then preventing and preparing for a potential crisis becomes much more difficult, if not impossible. We have to get these assessments right because they lay the foundation of everything else we do.

Emergency management's focus has changed over time as our collective risk has changed and, more accurately, as politicians' perception of risk has changed. During industrialization, the increase in industrial accidents created new types of risk to be concerned with, as the threat of nuclear war did later. Our last major shift in focus came in the wake of 9/11, as our emergency management system was forced away from an all-hazards approach and toward a focus on terrorism. At times these shifts have been attempts to focus on urgent risks we face but often they have been rooted in other factors—like politics—and have come at the expense of making us more vulnerable to other hazards.

The events we prepare for are not necessarily the events we are most at risk of experiencing. Despite our federal orientation for emergency management pointing toward homeland security and terrorism fewer than 1 percent of all federally declared emergencies and disasters have been terrorism related. By contrast, 99 percent of US counties have been impacted by flooding since 1996[445] (and 100 percent have been affected by a pandemic). Preparing for a terrorist attack is important, but such a disproportionate use of resources underscores how our preparedness priorities don't always align with our true risk.

From the beginning, there were signs that a Trump presidency would end in catastrophe. His first attempt at a disaster response was the flooding in Louisiana just months before the 2016 election. Presidential visits in the days following disaster often face criticism. If presidents wait too long they are accused of not caring, but if they visit too soon they can grind response to a halt as already limited resources are diverted away from relief efforts to the arrival of the president.[446]

This is why in 2016, Louisiana governor John Bel Edwards asked President Obama to delay his visit until the immediate response to the flooding had concluded.[447] Then-candidate Trump ignored this request and showed up in Louisiana without notifying the governor. In a prelude to the more jarring images of the president throwing paper towels at Puerto Ricans after Hurricane Maria, Trump stood in a crowd passing out packages of Play-Doh. Just as paper towels won't clean up after a hurricane, Play-Doh doesn't rebuild houses.

Once elected, Trump nominated Brock Long, who was among the least controversial of Trump's nominations, to run FEMA.[448] Long had served as the director of Alabama's emergency management agency and seemed to have a favorable reputation within the emergency management community. Emergency management is much more than FEMA though. Other agencies have a significant bearing on emergency management and their directors oversee policies and programs related to mitigation, preparedness, response, and recovery.

Appointments made within these other agencies were immediately troubling. John Kelly was made secretary of Homeland Security with what appeared to be only a superficial understanding of FEMA.[449] Scott Pruitt was nominated to lead the Environmental Protection Agency despite, and perhaps because of, his close ties to the oil industry.[450] Ben Carson, having no relevant experience, was made the Secretary of Housing and Urban Development.[451] Tom Price was nominated to lead the Department of Health and Human Services despite questions about his buying and selling of stocks in health companies as a member of Congress. He resigned within months anyway amidst a scandal involving his use of taxpayer dollars.[452] From the beginning, these nominations, and the negligence with which they were selected, signaled a challenging future for managing disasters.

Then there was the budget. In an effort to find billions of dollars to put toward the president's obsession with "the wall" along the southern border, the administration released a draft budget proposal in March of 2017. The proposal included cuts across federal agencies that would have been damaging to our national preparedness. FEMA's proposed cuts were particularly egregious.

The administration called for an overall 11 percent budget cut to FEMA at a time when the agency was already considered underfunded.[453] The *New York Times* reported the cuts specifically targeted "an array of grants to state and local governments that have helped fund the development of emergency preparedness and response plans." The proposal sent a shockwave through the emergency management community. One emergency manager in Ohio summarized the sentiment felt by many when he told the *Washington Post* they were "draconian cuts" that would be "damaging [to] the national system."[454]

The FEMA cuts paled in comparison to the downright hostility directed toward the EPA. The administration sought to cut the agency's budget by $2 billion including 69 percent of funding for climate change–related work and a 78 percent cut to environmental justice programs. Their plan was to fire a fifth of the EPA's staff and eliminate at least thirty-eight programs. The National Oceanic & Atmospheric Administration, the Department of Housing and Urban Development, the National Weather Service, AmeriCorps, the Coast Guard, and TSA all faced cuts of their own that would have, in various ways, undermined our emergency management system and made us less prepared. Though, in hindsight, perhaps it was the proposed $12.6 billion cut from Health and Human Services that is most chilling.[455]

Although most of these cuts were never fully implemented, each of these agencies experienced some level of defunding and/

or turmoil in the first year of the administration. The proposed budget gave us the first real glimpse of how seriously, or not seriously, this administration planned to take emergency management. As they say: follow the money. Their budget wasn't designed to prepare us for the greatest threats we face. If it had been, climate change and public health would have been prioritized. So, as the 2017 hurricane season began, this was the state of the federal government.

Throughout his tenure, President Trump created a spectacle of disaster response.[456] As Harvey made landfall on August 25, a reporter asked the president if he had a message for the people of Texas, as they faced one of the worst hurricanes in US history. He gave a thumbs-up and said, "Good luck to everybody."[457] It was a statement that did not inspire much confidence or empathy. Days later, Melania Trump boarded Air Force One for Texas wearing stilettos, which further underscored the lack of awareness of the devastation that awaited them in the South.

The spectacle continued days later as Irma cut through the Caribbean, affecting both Puerto Rico and the US Virgin Islands. After the storm, the president announced he had met with the *president* of the US Virgin Islands.[458] The misstatement (he met with the governor) fueled the impression that he was treating the response to the two US territories as foreign aid.[459] It raised concerns again about the lack of presidential awareness and sympathy to the areas affected by disaster and, worse, what communities could expect in terms of how actual distribution of federal aid would unfold.

This was a pattern throughout his administration. Each year after major wildfires in California, Trump would allude to or directly threaten to withhold federal aid from California.[460] For about forty-eight hours in October 2020, it looked like he had made good on his threat when he denied a declaration request from the governor of California. After public outrage and a phone call with House minority whip Kevin McCarthy, a Californian Republican, he reversed the decision and approved the declaration.[461] Back on the East Coast, in response to a proposal for a seawall to help protect New York City from flooding he Tweeted, "Sorry, you'll just have to get your mops & buckets ready!"[462]

These moves were certainly insensitive, but what the president needs to do operationally in disasters like these is sign the Presidential Disaster Declaration so federal support can be provided. To his credit, during disasters like Harvey and Irma Trump signed the declarations with little fanfare.[463] Texas and Florida got the resources they requested from the federal government, and while there were plenty of people, particularly marginalized groups, whose needs went unmet, the response to these disasters went about the way we would expect.

The exception was, of course, the failed response to Hurricane Maria, which went far beyond petty Tweets and inappropriate footwear. The gaffes and insensitivity escalated to inaction and hostility. The Maria response revealed both deep systemic problems with our emergency management system and also highlighted the cruelty of the Trump administration.

FEMA's vulnerability, due to its position within DHS, emerged throughout Trump's presidency as well. As Florence approached North and South Carolina, rumors of infighting between FEMA administrator Brock Long and the secretary of DHS Kirstjen Nielsen leaked. Nielsen reportedly disliked Long and threatened to fire him—in the middle of a major hurricane. The extent of damage that resulted from infighting is a point of speculation, but it certainly created a distraction, if not additional challenges, to the agency's response, and demonstrated the vulnerability of FEMA's authority in a crisis.[464]

Further, the vulnerability of FEMA's funding to the agenda of DHS was on full display. Twice during the Trump administration, FEMA had millions of dollars transferred from their budget. Congress had refused to fund the president's wall project, so he ordered his administration to find the money in existing budgets. As Hurricane Florence crawled toward the Atlantic coast, $9.75 million was moved from FEMA to ICE.[465] FEMA leadership gave assurances that the move would not jeopardize the hurricane response because the funding came from the agency's routine operating budget rather than the Disaster Relief Fund.[466] Of course, it is that routine operating budget that builds the capacity of the agency and allows them to do their work. The move may not have plucked a check out of the hand of a Hurricane Florence survivor, but it was a hit on the agency's ability to be ready to prepare for future responses. The following year it happened again. As Hurricane Dorian neared Puerto Rico and the East Coast, $271 million was taken from FEMA and given to ICE. This time the money was taken from the Disaster Relief Fund, the primary source of funding for FEMA's public and individual assistance programs.[467] These decisions were just the latest in the long history of FEMA's mistreatment.

The government has failed to build an emergency management system to meet our growing needs. FEMA is a shell of an agency with a meager and insufficient budget, perpetual understaffing, and minimal authority. They are kicked around at the pleasure of the president and used by Congress as a scapegoat. And at the state and local level a similar dynamic often plays out between state and local emergency management agencies, and state legislatures and local politicians.

The system that we have has been slowly built, piece by piece, disaster after disaster, for decades. You cannot build an effective response to a catastrophe in a day. It requires the use of well-tried systems and procedures that facilitate the collaboration, communication, and coordination among thousands of organizations. There has to be trust. When the pandemic arrived to the United States in 2020, it revealed to everyone else what disaster survivors and experts have been pleading with everyone to see: we aren't prepared.

How Not to Manage a Pandemic

GEORGE W. BUSH'S summer reading in 2005 was John Barry's book *The Great Influenza: The Story of the Deadliest Pandemic in History*.[468] Not exactly a beach read, the book outlines the history of the 1918 pandemic. After finishing it, Bush reportedly became concerned that America was unprepared to manage another such pandemic. He directed Fran Townsend, his top Homeland Security adviser, to create a national strategy immediately.

His advisers were hesitant to dedicate time and resources to the effort, given they were in the middle of managing the aftermath of their failed response to Hurricane Katrina. The president insisted, though, and by November Bush gave a speech at the National Institutes of Health in which he told them what they already knew: a pandemic would one day come to the United States, and we needed a plan. The Bush administration set out to spend $7 billion on a variety of plans, procedures, and exercises that, though not comprehensive, laid important groundwork for an eventual US pandemic response.

Experts from all over the world have long warned of the pandemic threat. The federal government was well aware of the risk too. As recently as July 2019, FEMA had identified a pandemic as a major threat to the United States. In their National Threat and Hazard Identification and Risk Assessment (THIRA), FEMA compiled a long list of well-known catastrophes like an earthquake along the New Madrid, San Andreas, or Cascadia fault lines and Category 5 hurricanes in major US cities.[469] One of these nightmare scenarios described a novel strain of influenza that would overwhelm the healthcare system and grind the country to a halt. By March 2020, FEMA's hypothetical pandemic scenario read like a historical account.

Other agencies across the federal government had also warned of a pandemic and the need to increase preparedness efforts. In the National BioDefense Strategy, released by the Trump administration in 2018, they used recent infectious disease outbreaks like Ebola and Zika as evidence of the need for countries to "improve their preparedness and biosurveillance systems to detect and respond to the next health crisis." To them, this was important because, as they wrote, "The health of the American people depends on our ability to stem infectious disease outbreaks at their source, wherever and however they occur."[470] The Worldwide Threat Assessment of the US Intelligence Committee, released in January 2019 by the Director of National Intelligence Daniel Coats, affirmed,

"We assess that the United States and the world will remain vulnerable to the next flu pandemic or largescale outbreak of a contagious disease that could lead to massive rates of death and disability, severely affect the world economy, strain international resources, and increase calls on the United States for support."[471]

Despite plentiful and consistent warnings spanning decades, Trump insisted the pandemic was "an unforeseen problem" that had "come out of nowhere."[472] While he may have been surprised by this threat, his administration, the country, and the rest of the world were not. He said that "nobody knew there would be a pandemic or epidemic of this proportion,"[473] a claim countered by the global scientific community, previous US presidents including those from his own party, his own advisers, and common sense.

Journalists like to ask me if such and such city is prepared for such and such disaster. Is New Orleans prepared for a Category 5 hurricane? Is California prepared for a massive earthquake? Is Saint Louis prepared for a major river flood? To be honest, I always sigh when they ask. The question itself reveals the limitations of how we think about preparedness.

Preparedness is not a "yes" or "no" question (though, if you make me choose, I will pick "no" every time). Preparedness isn't a static state, it's dynamic.[474] That's why, when I sat on the beach in Hawaii gaming out the different disaster visions, I knew we were headed for big trouble. Was the US prepared for the pandemic? Clearly the short answer is "no"; the longer answer, though, is more complicated. It is simultaneously true that we had prepared for a pandemic, and that we had not done enough to prepare for a pandemic.

In the footsteps of the Bush administration, and after the global response to Ebola, the Obama administration laid the groundwork for a relatively effective pandemic response. Obama aides even included a pandemic response scenario, eerily similar to the one that unfolded in 2020, in a transition meeting held with Trump's top aides to emphasize the risk of this threat.[475]

When Trump took office, the infrastructure for a national response to a pandemic was in place. It was not as well-trained or planned for as many would have liked. More could—and should—have been done. For example, it was long known that the Strategic National Stockpile and medical supply chains were not well prepared for a threat of this nature.[476] Still, there were experts in positions of authority and there were serious people who were concerned with this risk.

It is also true that from the moment Trump took office, his administration chipped away at our national preparedness. One of the more blatant moves to undermine our preparedness came in 2018 when they disbanded the pandemic response team, officially called the Global Health Security and Biodefense unit, that had been created within the National Security Council during the Obama administration following the response to Ebola.[477] Beth Cameron, the former senior director for the team, wrote in an op-ed that their mission had been "to do everything possible within the vast powers and resources of the US government to prepare for the next disease outbreak and prevent it from becoming an epidemic or pandemic."[478] Through actions like this one the Trump administration further depleted the systems and resources that were in place to manage such a crisis while installing science deniers and friends of industry throughout the federal government.

A more general, and perhaps fundamental problem for our national preparedness was that the White House had turned the federal government into a vat of toxic waste. The administration was in a constant state of controversy connected to hidden agendas, lies, backstabbing, hostility, lawsuits, and criminal investigations. Governments that are in a state of chaos on a good day will not all of a sudden function coherently during times of crisis—yet another lesson that might have been learned from the failed response to Maria if more attention had been paid.

Worst of all, the administration had eroded trust with the media and the public. While in office, the *Washington Post* found the president lied or made misleading statements tens of thousands of times.[479] Through his persistent claims that journalists and media outlets are "fake news" the president worked to undermine one of the most important resources in times of crisis.[480]

As the pandemic swept the country, it became clear that the Trump administration was unwilling to act in a way that would prevent mass death. In January, a group of public health experts began writing to each other with their concerns about the government's response to the inevitable arrival of COVID in the United States. The *New York Times* published a series of emails, called Red Dawn (after the 1984 film starring Patrick Swayze and Charlie Sheen) that demonstrated the escalating concern among experts.[481] On March 12, Dr. James Lawler of UNMC and former adviser to the Bush and Obama administrations wrote a damning assessment:

> "We are making every misstep leaders initially made in tabletops at the outset of pandemic planning in 2006. We had systematically addressed all of these and had a plan that would work—and has worked in Hong Kong/Singapore. We have thrown 15 years of institutional learning out the window and are making decisions based on intuition."

These experts put into words what many of us felt about the response to this pandemic. Despite the damage that had been done to our national preparedness, there was still a wealth of resources and expertise that could be called on to help guide the response.

Instead, they inexplicably sent Jared Kushner to FEMA to lead PPE procurement efforts.[482] This decision was perplexing as the president's son-in-law had no prior emergency management experience.[483] Despite the nondisclosure agreements they were told to sign, some involved with Kushner's efforts have spoken out. What they describe is a bizarre, and unnerving, approach to gathering the supplies needed to protect healthcare workers around the country.

In a conference room at FEMA headquarters a group of seemingly random twenty-year-old *volunteers* were gathered and told to find PPE. They claim to have been given no instruction and did not even know the rules of how government procurement worked and were left to Google leads of possible stockpiles of masks and other PPE around the world. They describe wandering around FEMA headquarters looking for help.

The administration chose to improvise the response over the recommendations of public health and emergency management experts. In spite of everything we knew about the Trump administration, it was still astonishing. There is a long tradition of successful improvisation in disaster response—from the Cajun Navy in New Orleans to the evacuation of Manhattan on 9/11—but this was improvisation that didn't seem to have the goal of saving lives. Rather the White House—and many governors—seemed to be in a race to see how quickly they could spread COVID around the country.[484]

In an interview in April, Dr. Kathleen Tierney put into words what had left so many disaster experts, including myself, speechless.

"This is the worst crisis response debacle I've seen in my career as a disaster researcher. The wheels have fallen off and everything it seems that can go wrong is going wrong. There was such a massive failure of foresight and a massive failure of execution and failure, frankly, to follow all of the planning philosophies and all of the plans and institutional arrangements that this country has been working on in emergency management for 50 years. I mean it's astounding. The incompetence is astounding."[485]

I have found myself repeating again and again since the pandemic began that this didn't have to happen. We know better than this. This was preventable. Disasters, and catastrophes, are a choice. They are a political decision.

There was an earth-shattering moment when I realized not

having a plan was their plan. The plan was to let the virus run its course and bank on a vaccine. Their plan was to keep the economy open regardless of the human toll. Their plan was to ignore the science. To pretend that hundreds of thousands of avoidable deaths was normal. They pretended that we should not be scared of a disease that kills when, in fact, you absolutely should be. It was then that the many possible scenarios I had imagined in February and March dwindled to just the most extreme, and the extent of death began to sink in.

Sometime between April 7 and 8 the official COVID-19 death toll in the United States officially passed over twelve-thousand deaths. The number hit me hard because it was already at the high end of the estimated death toll of the 1900 Galveston hurricane, meaning we were out of traditional disasters for which to compare the death toll. The *New York Times* had sent out a push notification when the death toll passed the number of people—three thousand—who had died on 9/11. We got another one when three thousand people started dying *each day*.

By Christmas 2020, scrolling my Facebook newsfeed felt like reading the obituary section of a newspaper. I watched the number of cases spike in New Orleans, a city where the same history that led to Katrina led to higher rates of chronic illnesses. I watched as locally elected officials in Houston fought back against Texas governor Abbott for a mask mandate—the simplest of actions to keep people safe.[486] I watched the doctors at UNMC, the pinnacle of public health, warn that they were on their way to being overwhelmed.[487] I watched an all-out bloodbath unfold across North Dakota where one in every eight hundred residents died after state officials did nothing to prevent the spread (as of this writing).[488]

The contrast in outcomes throughout the various states based just on leadership and basic public health guidance could not have been more stark. In Maine, Governor Janet Mills took the pandemic seriously. She issued mask mandates, put more money toward unemployment benefits, worked to build our hospital capacity, and deferred to the state's CDC director Dr. Nirav Shah. As a result, the state's case numbers remained among the lowest in the country for most of the summer of 2020.[489]

A report from the National Center for Disaster Preparedness at Columbia University estimated that as of mid-October 2020 between 130,000 and 210,000 deaths could have been prevented in the United States if different policy interventions had been made early on.[490] I thought of the quote late folk singer Utah Phillips once said, "The Earth is not dying, it is being killed, and those who are killing it have names and addresses." People didn't die of COVID, they were killed. COVID was perhaps inevitable but the way we responded to it was a choice—just as how we respond to hurricanes, tornadoes, and all the other known hazards is a choice. The violence of the pandemic, the despair, the pain, the death, were decisions made by people who have names and addresses.

In an interview with Bob Woodward in March, Trump admitted he was intentionally downplaying the situation to prevent panic.[491] The day the tapes of the interviews were released, I thought of Dr. Quarantelli and all the other disaster researchers who for decades had fought against this disaster myth. The public rarely panics in a disaster. Leaders have to communicate clearly and accurately in a crisis. They have to tell the truth. They need to give the public information as they get it so the public can make informed decisions to protect themselves and each other.

The silence of January and February was met by the agonizing and confusing rollout of shelter-in-place orders. We learned a whole new vocabulary in March. Flattening the curve, social distancing, testing, COVID, and quarantine permeated conversations and affected every action. When elements of the pandemic response were successful, it was because of the competence of scientists, investigative and local journalism, activist networks, the leadership of some state and local officials, and the good-will of those of us who followed them instead of the advice of the Trump administration. Mutual aid networks sprang up in every city across the county—there were homemade sign-ups to get groceries for the elderly, children's authors livestreaming story time for kids, volunteers sewing masks, and crowdfund-ing for people who needed financial help. Meanwhile words changed meanings as *essential* came to mean *expendable*. States fell like dominoes across the country as hot spots emerged and drained the US emergency management system.

By the end of February, Fox News and the president were still calling the need to respond to the coronavirus a hoax.[492] Meanwhile experts recommended necessary actions hoping to "flatten the curve." Staying home was only a temporary stop-gap, though, an extreme effort to prevent a tremendous loss of life akin to the 1918 pandemic, during which it is estimated that at least 675,000 Americans died.[493] Flattening the curve would still mean tens of thousands of deaths, but it meant that hun-dreds of thousands of lives could be saved.

Although the White House led the public to believe that flattening the curve would be temporary, public health experts voiced more dire warnings. The need for stay-at-home orders and social distancing could not be short-term without risking another spike. For those making responsible decisions, Easter would not be held in churches. It would not only be the spring semester that was moved online. We would not just be home-bound for a few weeks. Those who were thinking ahead knew Thanksgiving and the winter holidays wouldn't be happening as usual either.

We have a saying in emergency management that "disasters begin and end locally." What we mean is that it's the local com-munity, the survivors of the disaster, that will be there first in response, and last as recovery ends. This is true, but the very nature of disasters—and especially catastrophes—is that the local community becomes overwhelmed. Disasters may begin and end locally, but it's the part in the middle that we need outside resources—including the federal government's—to help us with.

The very nature of the pandemic—the scope and scale—demanded a coordinated national response. We don't have border walls between states—they are not islands unto themselves, and what happens in one affects the others. COVID was not going to abide by jurisdictional lines. It turns out the whole "united we stand" thing they had drilled into our heads after 9/11 didn't apply to this pandemic. As the White House encouraged governors to not implement basic public health measures, like mask mandates, it became clear they had chosen "divided we fall."

And we fell hard. People lost their jobs in waves, businesses failed, there was no assurance of government assistance in any kind of substantial or sustained way. The disparities in impacts were revealed quickly.[494] While everyone was affected by the pandemic in some way, the severity of those impacts was not equal. People of color, the elderly, and the chronically ill died at disproportionate rates.[495] Even in places like Maine where the response to the pandemic was handled more effectively (as compared to states like Florida or North Dakota) the disparities were still evident. By the end of June, Maine had the highest racial disparity in confirmed cases of any other state.[496]

While many in the middle class were able to continue working from the safety of their homes, the working poor were given the choice between risking their lives at work or not having an income. As elected officials continually failed to provide adequate financial relief to the public more lives were put at risk. In Las Vegas, in the shadow of empty luxury hotels, the city drew squares on the ground of a parking lot for the homeless to sleep in.[497] When the president was flown to Walter Reed, we watched the wealthy buy themselves a cure and months later watched as they tried to buy the vaccine for thousands of dollars.[498]

As much as the pandemic would require a nationally coordinated response, it would also require each state to coordinate their own response, meaning there would be fifty-plus responses occurring simultaneously across the country, for months on end. Never before in US history had every single emergency management agency been activated at once, but that was exactly what the pandemic required.

Our approach to emergency management relies on help coming into affected communities from outside, unaffected communities. As COVID spread across the country I worried about where that help would come from—as strained as the emergency management system was in 2017, the COVID response was something else entirely. Even the most competent administration would have had difficulties in managing the kind of complex response required for this situation.

While many state and local emergency management agencies across the country were trying to use preexisting systems and plans that had been in place, it was harder to understand the approach being taken by the federal government. One egregious example was related to the individual assistance program overseen by FEMA.

On March 13, 2020 President Trump declared both a National Emergency (through the National Emergencies Act) and a State of Emergency (through the Stafford Act). For emergency management it was the use of the Stafford Act that was most interesting. The way in which the Stafford Act is written is a bit vague and sometimes there are questions about the applicability of the Act to a given crisis.

When the pandemic began and the idea of using the Stafford Act as a relief mechanism was raised, some questioned if it could be used for a pandemic. In fact, then-senator Kamala Harris even introduced a bill in Congress to amend the Stafford Act to explicitly include pandemics, although it was never voted on.[499] Former governor Tom Ridge, the first secretary to the Department of Homeland Security, was also one of the main authors of the Stafford Act. In March he wrote an op-ed confirming what those in emergency management already knew, that even without the word *pandemic* explicitly articulated, the intent of the Stafford Act was that it should absolutely be used in a pandemic response.[500] If you're not going to use the cornerstone of emergency management policy in the midst of a catastrophic pandemic, I don't know when you would!

Although President Trump made a blanket emergency declaration for the whole country, it was really a major disaster declaration, through the Stafford Act, that was most needed. This kind of declaration would require the governor of each state to request a declaration from the White House. It took weeks but eventually every state and territory requested and received a disaster declaration, which enabled a better flow of federal resources specifically for public assistance.

The odd thing was that a number of states had also requested individual assistance—the kind of aid that homeowners receive to help rebuild their houses in the wake of flooding. While the pandemic didn't cause physical damage in the way that is typical during a disaster, individuals across the country did have needs that could be addressed through existing FEMA programs. Yet, the White House did not approve the use of these programs.

One of the obvious sources of aid that went unused was FEMA's funeral assistance program.[501] In the wake of a presidentially declared disaster victims' families may be eligible for financial assistance to help cover the cost of funerals. It's not much money but it can be a big help during a difficult time. Despite this program being requested by many governors, the White House did not approve its use. Instead I watched for months as people—including my friends—who had lost a family member to COVID made GoFundMe pages to help cover funeral costs.

This existing federal program, which was created for this exact situation and that the president could approve the use of on his own, without Congress, was just sitting there unused even though there was plenty of money in the Disaster Relief Fund to cover the cost. What was so frustrating about the federal response was that even though there were inadequacies in how we had prepared for a pandemic there were existing tools and programs, like funeral assistance, that weren't being used at the time people needed them most.

FEMA faltered, as you would expect from an agency that had been systematically weakened and then told to help coordinate a response to a crisis they were never prepared to manage.

Our emergency management system had never been built for such a response. Every day I heard from very tired—and angry—emergency managers across the country, at all levels of government, who were trying to move mountains to help their communities. They were throwing all the resources they had at the response, but there was only so much that can be done to make up for decisions being made by elected officials that were seemingly designed to encourage the pandemic to spread. As early as April, I heard emergency managers using terms like the *verge of collapse* to describe the state of local emergency management and public health agencies. I learned of the deaths of emergency managers across the country who had worked through the pandemic. They had been pushed into a machine of misinformation and local political frenzies, and had no armor.

This response was a spectacular failure, but it should not have come as a surprise. A group of people who do not know anything about managing a catastrophe, who do not understand the basic functions of the federal government, and who lack compassion, cannot successfully manage an unprecedented crisis, particularly after working to dismantle the nation's public health and emergency management systems.

President Bush may also have been enlightened by one of John Barry's other books in the summer of 2005, *Rising Tide:*

The Great Mississippi Flood of 1927 and How It Changed America.[502]
The book outlines another famous levee failure in New Orleans
and its repercussions. The story of the 1927 flood also lends in-
sight into Trump's response to COVID. While Hoover was in
the Midwest visiting Red Cross relief camps, business leaders
in New Orleans watched as the flooding made its way south.
Unlike the piecemeal approach to levee building that had been
used throughout the Midwest, the city of New Orleans had already built a more coordinated levee system along the river. Although it was believed the levees would hold in New Orleans
(because breaks upriver would release the pressure) there were
concerns about the economic impact on the city even if there
was no significant flooding.

The *perception* that New Orleans could flood was enough to
threaten business interests in the city. So the city's business leaders pressured officials into blowing up the levee to ensure the city
would not flood. And so state engineers dynamited the levees at
Poydras. Little warning was given to residents, as they rushed to
gather their belongings before their homes were flooded. The
people who lived and worked in the flooded area didn't have
the political or economic power to fight back. Further, despite
promises to the contrary, they were never adequately reimbursed.

To be clear, dynamiting the levee was done for the benefit of
economic interests, not an actual threat of flooding. The city's
elite fabricated a disaster in an effort to obscure the actual risk
to the city, while creating a financial opportunity for themselves, and displacing marginalized residents. In other words,
elite business interests were prioritized over the well-being of
the public. It was, in a word, sinister.

The Trump administration blew up the response to COVID
in what was presumably a misguided effort to save the economy.
They ignored the system that was in place to protect us, all in
the service of our supposed financial interest. Few seemed to
think twice as they sacrificed the marginalized and vulnerable.

The federal government did not even bother to adequately compensate them. The complete disregard for human life demonstrated by many elected officials and even some in the public is terrifying in its own right, but when you apply what has happened in the pandemic to the context of climate change, you can see why we have to change everything. In fact, the only thing that ever surprised me about the COVID pandemic was that we were not *all* out in the streets, demanding the federal government provide assistance and an effective response.

PART SIX

Disaster Justice

"I wanna help."

ONE SUMMER, ON a nonpandemic drive with my brother from North Dakota to Maine, I suggested we take a detour a bit farther north to Niagara Falls, New York. My destination of interest wasn't the falls but rather an old abandoned neighborhood. We arrived to find overgrown sidewalks and streets. Plants had wound their way up and around fire hydrants and there were driveways that led to nothing. The neighborhood was quiet and reminded me of the long un-rebuilt streets in the Lower Ninth Ward.

In the 1950s, this was a newly built neighborhood called Love Canal and it was filled with modest, working-class homes.[503] As the neighborhood grew the board of education purchased a piece of land on which to build an elementary school. By 1978, there were over a thousand families living there. They sent their kids to school, walked their dogs, breathed the air, and drank the water. The new residents did not know it then, but they were living on top of one hundred thousand barrels of toxic chemicals.

The families who had moved into Love Canal were not told

that the Hooker Chemical and Plastics Corporation had spent years dumping toxic chemicals into the old canal. When they sold the property to the board of education in 1953, Hooker did not remove the waste, instead just piling new dirt over the site. They also included a release of liability for any illnesses and deaths connected to the site. The board of education went ahead and built the school and the city continued to allow developers to build what became the new neighborhood, around the edge of the contaminated property.

Local residents had long suspected something was not right. Around the time Hooker first began dumping chemicals, residents reported strong smells and spontaneous fires that broke out in hot weather. By the late 1970s, they began documenting high rates of illnesses in the neighborhood, particularly among children. Residents experienced various health ailments for many years but it wasn't until an increase in coverage in the *Niagara Gazette* that the connections to the chemical site became understood. Scientists confirmed the chemicals had seeped through the ground and into people's homes and yards. A number of studies found a disproportionately high rate of birth defects, increased rates of cancer, and other medical conditions in the neighborhood.

For years, there was an underwhelming response from government. In fact, the mayor and other local officials actively ignored and avoided the budding crisis. As media coverage increased and the truth came to light, residents began to organize through the Love Canal Homeowner's Association.[504] Their goal was to elicit action—namely for government to buy out the homes in the affected area so residents could move. It took years, research, national media coverage, and political negotiations between the state of New York and the federal government, but residents in the immediate area were eventually given their buy-out and efforts were undertaken to clean up the site. It took two decades and $350 million to address the Love Canal crisis.

Policy doesn't just change on its own. It requires advocacy and organizing to hold elected officials accountable. Across the country people—often led by women—have done this work. A woman named Hazel Johnson rallied the concept of environmental justice after living the effects of environmental racism in the South Side of Chicago. Her work led to President Bill Clinton signing the first major federal action explicitly related to environmental justice.[505] After the buy-out the leading activist in the Love Canal crisis, Lois Gibbs, continued on to lobby the federal government to create a program to provide funding and assistance to communities across the country facing similar crises. Standing on the foundation laid by the work of others like environmentalist Rachel Carson, advocates were able to capitalize on the national coverage of Love Canal and push for changes to federal policy. Through Gibbs's and others' efforts, the EPA created the Comprehensive Environmental Response, Compensation, and Liability Act, known as the Superfund Act, which provides funding and resources to communities who need to clean up toxic sites.

It was the EPA's Superfund program, championed by the survivors of Love Canal, that was used to clean up the CIPS Gas Plant site in my hometown of Taylorville decades later. Standing in this now-abandoned neighborhood in New York I felt an unyielding gratitude for the people who had lived there decades earlier. Without their efforts the small town, across the country, that I was born into would not have had a road map for how to resolve their crisis.

Creating a Movement for Disaster Justice

IN OCTOBER OF 2016 we were quickly nearing the months when Fargonians hibernate for the winter. I was in the middle of the stressful process of writing up the findings from my research trips to Texas for my dissertation. A friend from college, one of the girls I had evacuated New Orleans with during Hurricane Gustav, decided she needed to stage a small rescue mission to get me out of the house before winter set in.

We planned a spontaneous weekend road trip out West to make the best of the last bit of warmer weather. My friend drove in from Minneapolis, picked me up in Fargo, and we headed west to Bismarck. From the capital we turned south, bound for the Badlands. As we made our way toward South Dakota, along an otherwise barren route, we were brought to a stop by concrete barriers and the North Dakota National Guard with guns in hand.[506]

In 2014, Dakota Access announced they would build a pipeline to carry oil from the Bakken oil fields in North Dakota to southern Illinois, just an hour south of where I had been born in Taylorville.[507] The pipeline was originally planned to cut through the predominately white city of Bismarck but was later rerouted along the Standing Rock Reservation through ancestral burial grounds and under the Missouri River.[508] The new route threatened to contaminate the drinking water supply of eighteen million people.

The legal fight began immediately, with groups in multiple states suing to prevent the pipeline construction.[509] In North Dakota, members of the Standing Rock Sioux Tribe explained how the pipeline would threaten their right to clean water and their way of life. Despite arguing that their sovereignty—as recognized by the US Congress in the Fort Laramie Treaty of 1851—was being infringed upon, the Corps of Engineers and other government agencies allowed the project to move forward.

What we had accidentally stumbled upon was the Standing Rock protest[510] that had begun months earlier with a prayer camp established by a group of Indigenous youth activists.[511] I was marginally aware of what was happening with the pipeline, but it had not yet received the mainstream coverage it would just a few weeks later. Despite my geographic proximity and concern for environmental issues, I had been focused on other things like finishing my dissertation in time to graduate. It was not until we stumbled upon this unnerving checkpoint that I understood this was not just a normal pipeline protest.

As you might expect, two young white women were not perceived to be a security threat and, without question, we were waved through the barriers. We drove on and passed a fence covered in flags that ran the length of the road. While I hadn't grasped the gravity of what was happening at Standing Rock, others from around the world certainly had. People from other Tribal Nations, countries, and organizations had sent flags in a display of solidarity, which had been hung end to end for as far as we could see.

A ways down the road, the camp came into view. There were rows of tents, tepees, and campers—a distinctive site in an otherwise isolated location. It felt like the protest was very contained. Not wanting to intrude, we carried on to South Dakota, with the newfound realization that this protest was bigger and more complicated than we had understood.[512]

I spent the next month learning about the Water Protectors from Fargo.[513] I more closely followed and kept up with updates from protest leaders like LaDonna Brave Bull Allard.[514] I watched the increasing media coverage from journalists like Amy Goodman and learned the history of the area and the treaties. By mid-November, the situation was escalating rapidly. The national media did not drop everything to cover the protests, but there was a growing awareness around the country as politicians like Senator Bernie Sanders and celebrities like Shailene Woodley and Mark Ruffalo raised their profile.[515] The mainstream media that did arrive to cover the protests seemed to not fully know what to make of the situation. When asked why the federal government was not honoring the treaties, a CNN reporter shrugged and said, "You got me!"

By Thanksgiving I was sitting at my parents' house watching a livestream of North Dakota police blasting peaceful Water Protectors with water cannons in frigid twenty-degree weather.[516] Someone was able to maintain a six-plus hour livestream on Facebook and showed the world rubber bullets and tear gas shot indiscriminately into a crowd of peaceful protesters. Twenty-six protesters were hospitalized and over three-hundred injured.[517] As a human being, I was disgusted by the government's violent actions. As a disasterologist, I was furious.

The governor of North Dakota requested help through the Emergency Management Assistance Compact (EMAC)—which had been put in place for the sharing of resources during emergencies and disasters—to bring in law enforcement agencies from surrounding states.[518] Over fifty law enforcement agencies from as far away as Louisiana and Maryland deployed to North Dakota. We were not only watching a livestream of police brutality; we were watching a livestream of police brutality facilitated by the emergency management system. Specifically, a tool of this system—mutual aid agreements—was being used to physically attack the very people they should have been protecting.

It is tempting, perhaps, to look at the situation in North Dakota and think that the emergency management system, or at least the tools of the system, had been corrupted. The history, though, of how the system was built and who built it suggests that it was operating in the way it had been intended—to serve white, privileged communities at the expense of marginalized communities. In an echo of the Black Lives Matter movement, the conflict at Standing Rock again raised the question of who government, especially law enforcement, are "protecting and serving." The willingness to protect white lives and property, to serve some communities but not others, is at the root of emergency management's persistent and biggest failures.

The turning point came just before Thanksgiving (the timing of which was not lost on many). The protests grew rapidly with over two hundred Tribal Nations standing in solidarity and sister protests breaking out in cities around the country. Tribal elders and protest leaders invited non-Indigenous people to join the efforts. Environmental activists, human rights advocates, and veterans, among others, came to North Dakota by the thousands. In December, the North Dakota Human Rights Coalition was invited to Sacred Stone to participate in an interfaith service, and I tagged along.

I went with a van full of people, this time having a much better idea of the situation at hand. The seasons had changed and the ground was covered in a dusting of snow that drifted to form little silos, foreshadowing the spires of the Black Hills farther in the distance. As we made the turn south in Bismarck, I stared out at the familiar sights of small farms, old churches, and fields of cattle waiting in anticipation for the unusual sight of Sacred Stone.

When it came into view, I was shocked by the growth of the protest site in just two months. The photos and video footage had not captured its sprawl. The protest no longer felt contained. The tents and tepees had multiplied and were joined by more makeshift structures, painted buses, and cars laid out end to end, all the way to the river and hills. This time there were no roadblocks or men with guns to stop us—they had moved into the hills above to watch.

On this particular day, an estimated eight thousand people gathered amid rising smoke under a clear winter sky. In below-zero temperatures, gusts of wind churned with the constant hover of helicopters circling in the distance. Food was cooked over open fires, and booths selling crafts and supplies lined the muddy snow-covered aisles. Veterans shoveled ditches and built tents.

I stood in a crowd with people from different places and of different faiths listening to speeches. It was freezing but being packed into the crowd brought some semblance of warmth. I was too short to see the speakers so instead I looked up at a clear blue winter sky and the sun trying its best to reach us. There were drones and helicopters circling above and a flag with a picture of earth on it was flying overhead. I watched it flapping in the breeze.

I heard Dr. Cornel West before I even knew he was there to speak. His distinctive voice carried over the crowd and under the earth flag as he contextualized the fight at Standing Rock among historical uprisings in the US. He connected the movement at Standing Rock to the Black Freedom Movement, reminding the crowd not to "let anybody tell you that the original sin of the United States was the enslavement of Black people. That was the second one. The first one was the mistreatment and the dispossession of land and bodies and murder and mayhem of Indigenous brothers and sisters." Standing Rock was a continuation of the civil rights movement and the precursor to the protests and marches that continued after the 2016 presidential election.

His words made me think more about the intersecting movements that were represented at Standing Rock. People from radically different places who had spent their lives fighting their own battles dropped everything to sleep in tents in the middle of a North Dakota winter. (Do you know how cold North Dakota is in the winter?!) Standing Rock married the right to clean water, the militarization and brutality of law enforcement, Indigenous rights, and climate change.

I stood on muddy packed snow until I could feel the cold sink in through the bottom of my boots. The sun began to set as we piled back into the van. As we drove back to Fargo, we passed a line of cars and buses carrying in reinforcements, most of them veterans. In Bismarck our cell service came back, and we saw on the news the easement had been denied. The Corps would have to conduct a full Environmental Impact Assessment, which meant the pipeline construction would be temporarily stopped.[519] It was a fleeting victory.

One of Trump's first executive orders was to expedite construction, and the pipeline was completed in 2017. The protests did not stop the pipeline, but for a moment, the arrival of thousands of veterans to stand alongside Indigenous peoples on the plains of North Dakota after months of blizzards, violence, and well-wishings from around the world, was enough to make the country stop and look. Standing Rock meant different things to different people, but for me it was a glimpse of how we fix this. Intersectional activism is the only way we survive the climate crisis and Standing Rock was, to me, evidence that it was possible.

I've been to all manner of protests in my life from Black Lives Matter to climate change. I went to my first protest before I could talk. I've protested BP on the Louisiana Coast and housing discrimination in New Orleans. I counter protested Westboro and antiabortion demonstrations in conservative Louisiana towns. I was in Athens, Greece, during the 2015 anti-austerity protests and at the Women's March in Washington, DC. Standing Rock felt different to me. I had never seen so many different people and movements come together in such a committed way. Despite the long history of Indigenous resistance, the frozen plains of North Dakota struck me as an unlikely place for the climate movement, so often centered along the coasts and in big cities, to blossom.

I have seen the ineffective approach of the emergency management system in Camp Ellis. I have seen it sputter in Houston and fail in New Orleans. I saw emergency management manipulated into a tool for state-sanctioned violence during the Standing Rock protests. Emergency management, which *should* have the purpose of keeping communities safe, is persistently failing to do so, not least because the people who are in positions of power are not confronting racism, capitalism, and the other manifestations of power that drive inequality in risk, impacts, and recovery.

The use of water cannons in freezing temperatures on peaceful protestors is not separate from the failure to prioritize rebuilding the Lower Ninth Ward. The catastrophe in New Orleans and the crisis in North Dakota are connected to Houston's development-driven policy that is drowning the city, which is connected to the oil-covered wetlands in southeast Louisiana, and President Trump throwing paper towels at Puerto Ricans.

In each of these places I have found people working to expose and right the injustices in this system. There are examples from all around the country of communities coming together to protect themselves and fight for their future. In Camp Ellis, I saw community meetings packed with residents. I saw organizers pushing for flood mitigation to be put on the ballot in Houston and around every corner in New Orleans is a small but determined community organization that rebuilt the city.

Despite persistent local efforts on local issues across the country, there has never been enough public pressure to make comprehensive national changes. The tradition thus far has been that, to the extent that there is disaster activism, it occurs in silos, largely separated by either disaster or place. After Katrina, New Orleanians stirred up public pressure for changes to federal policy, as did the East Coast following Sandy, but the efforts largely faded. Even efforts related to affected communities in Texas, Florida, and Puerto Rico after the 2017 hurricane season were largely separate endeavors.

At Standing Rock, I saw clearly the connections between various movements across the country. In the years since, the whispers of a national "disaster movement" have grown louder. The Standing Rock protests, the public outcry over water quality in Flint, Michigan, Hurricane Maria survivors rallying in the streets of Puerto Rico and Washington, DC, the Sunrise Movement and other youth-led climate groups, and the March for Our Lives are all products of crisis.

These movements and protests, the majority youth-led, are in large part a response to disasters, and demonstrate the urgency and righteousness of the broader movements they represent. Disasters are themselves a manifestation of why these movements matter. Each protest intertwines the climate change and environmental justice movements, anti-austerity, anti-colonialism, anti-racism, and gender equality. Disaster activism is built from the foundation and guidance of these other social movements. Disaster is a common thread that weaves these movements together. If that thread was recognized and elevated, a unified disaster movement could emerge.

When I left New Orleans for Fargo, I thought I was going to graduate school to learn how to more effectively manage volunteers and donations after a hurricane. I thought I'd learn about a handful of policies that needed to be implemented to allow communities to recover more quickly. In the years since, I have realized that our problems are so much bigger. There is no single policy, no magic amount of money, no perfect plan that is going to meet the needs of communities across the country. It is much more complicated than that. Our failure to meet the needs of these communities requires no less than a movement for disaster justice. I have often found myself wondering what such a movement could look like. I wonder what it could do—I wonder what *we* could do.

There are absolutely people around the country (and world) who are already doing this work, but the efforts in one place are not being connected back to the work being done in other places in any strategic way. And so, in the absence of a broader movement, it is easy not to recognize our power—and we *do* have power.

Our power comes from our shared experiences of injustice—done to us or witnessed by us. Now we need to transform those experiences into organized action. And, like Lois Gibbs and the Love Canal activists, we cannot stop once our own crisis is resolved. We must push forward to protect communities everywhere. People will go to great lengths to protect themselves, their land, and their livelihoods—something that those in power have a history of forgetting.

Working Toward Disaster Justice

ONCE YOU KNOW the history of how our emergency management system came to be, how it was initially built of leftover policy from the era of civil defense and then disrupted by post-9/11 knee-jerk reactions, the inadequacies of how we do emergency management make a lot more sense. This cobbled-together, underfunded system isn't working. This is particularly troubling in this moment—the dawn of the climate crisis—as we face more severe natural hazards and other climate change impacts on our communities. We can lessen the pain of the climate crisis but doing so requires immediate and comprehensive emergency management reform on a scale that surpasses any since disaster policy was first written.

In any country that takes the safety of its residents seriously, an effective emergency management system is the bare minimum. Yet, thus far, the majority of our politicians have refused to make the very basic changes that could make the country better able to manage disasters. They have not learned the lessons of past disasters and we are living with the consequences.

We need our elected officials to act. How disasters are created, how we manage them, and what we do in their aftermath are all ultimately political decisions. When elected officials say during disasters that now isn't the time to talk about climate change or to talk about policy changes—they are gaslighting us.

Politicians, including presidents, have always used disasters to further their agendas. As the saying goes, "never let a good crisis go to waste." The political elite in this country have no problem taking advantage of disasters for their own gain. If they did, Herbert Hoover never would have been president and the Department of Homeland Security would not exist. When we allow politicians to treat disasters as though they are not political and when we don't advocate for change, we give away our power.

Despite the flaws and inadequacies of our current approach, something is still better than nothing. The emergency management system that we do have can't be taken for granted. It is not in the Constitution that FEMA will continue to help with disaster recovery or that Congress will keep reauthorizing the National Flood Insurance Program. We have to fight to keep these programs while we fight to make them better.

Climate change has demonstrated, perhaps more urgently than any other issue, the power structures that exist across the world and within the United States. It is no accident that the people who will and are most affected by disasters and the climate crisis, marginalized communities, are often the same people whose votes are suppressed.[520] Marginalized people will bear the brunt of the climate crisis. If you are able to vote, you need to consider how that power can be used to benefit frontline communities. As Stacey Abrams says, "We must unite our laws with our values." If we value safety, then we must address the consequences of climate change. If we value justice, then we must take a just approach to managing these consequences, including the growing number of climate-related disasters.

Making the perfect preparedness kit and volunteering post-disaster aren't going to be what keeps us safe in the long run. We need the kind of change that comes through organizing.

Educate

Reading this book (and the endnotes) is an excellent start, but this is just the beginning of the education you will need to enact change. You'll need a more nuanced awareness of disaster history; an operational knowledge of emergency management agencies, funding, and programs; a working understanding of environmental justice; and an ability to put this knowledge in a local context.

Some Book Recommendations to Get You Started

- *A Paradise Built in Hell: The Extraordinary Communities That Arise in Disaster* by Rebecca Solnit
- *Extreme Cities: The Peril and Promise of Urban Life in the Age of Climate Change* by Ashley Dawson
- *This Changes Everything: Capitalism vs. The Climate* by Naomi Klein
- *All We Can Save: Truth, Courage, and Solutions for the Climate Crisis* edited by Ayana Elizabeth Johnson and Katharine K. Wilkinson
- *Environmental Justice in a Moment of Danger* by Julie Sze

Every town, city, state, Tribal government, and territory (and country!) is different, so you need to consider how the issues raised in this book manifest where you live. Even within a community there are important differences that need to be understood. Learn about the parts of your town that have particularly high risk and the specific barriers that your community faces to implement mitigation and preparedness efforts. You'll need to educate yourself on these nuances.

What you have learned in this book has hopefully given you an idea of what the field of emergency management looks like and introduced you to the language needed to be able to do more research on your own.

Organize

Education is a lifelong continuous process but, at some point, if you want to make a difference you've got to do more than just go to book clubs. You need to make a plan. You need to strategize. What are the most pressing issues for your community? What is the problem you want to address? Why is this problem important to you? Why is this the problem *you* should address? How much time are you able to devote to solving this problem? Why should other people care about this? What is the history of the problem? What are the solutions? What are the steps to achieve these solutions?

Here are just some of the kinds of issues that need more public involvement to look for in your own community:

- Stop a developer from building new houses in a floodplain.
- Hold the Corps of Engineers accountable for addressing hazard mitigation needs in your community.
- Encourage state/local government to establish a buy-out program or prevent government from forcing a buy-out program on members of your community.
- Encourage your city to update drainage systems to prevent regular street flooding.
- Advocate for an increase in mitigation and preparedness funding locally and/or nationally.
- Call for the removal of FEMA from DHS.
- Demand emergency management policy be included in climate plans that address adaptation.

When you are ready to act, look around for the people already doing this work. Are there already people in your town advocating for a buy-out program for high-risk properties? Is there already a group working on wetland restoration? Across the country there are nonprofits and informal groups working on all manner of disaster mitigation and preparedness efforts. Most are not that big, and even if you live in the town where they work, you may have not heard about them before so start Googling!

When you find a group you're interested in, reach out and ask how you can join or support their efforts. This might mean offering financial support or volunteering your time. You may also have a special skill set, like graphic design or legal expertise, that could be particularly helpful to the group's efforts.

If you cannot find any groups that are specifically working to address the specific problem you have identified in your community you might want to consider checking in with your local climate, environmental, housing, or other organizations and mutual aid groups focused on social issues anyway. They may not be aware of the problem you want to work on, but because it may be related to the issues they work on they may be interested in working together.

Power does come in numbers but you do not need to be a part of a group to do disaster advocacy. If you have no luck with existing groups, you can always start your own. Make a hashtag, recruit your friends and family, and start talking to people in your community about whatever problem you're working to address. It may take a while but people will eventually join you. It is also a good idea to connect with other groups around the country that are doing similar work.

Take Action

Whether you are in a group or working on your own, there are
plenty of tactics you can use to hold elected officials accountable when it comes to emergency management.

Show up to government meetings. Budget meetings, planning hearings, and even disaster-specific meetings like the ones held in
Saco about Camp Ellis are often not well attended by local residents. Look into when these meetings are and what will be
discussed. Usually there is a LISTSERV you can sign up for or
social media pages that can notify you of upcoming meetings.
If you cannot attend these meetings in person check to see if
there is a livestream. At the very least, their minutes should be
posted online afterward.

Yes, these types of meetings can be very boring but often there
aren't many people there, so it is a good opportunity to advocate
for your problem to the people who are in a position to make
changes. (FEMA sometimes recommends that emergency managers bring cookies to hazard mitigation planning meetings, so
you might even get a snack out of it.) It may seem like advocating for changes at a local level is too small an action to make a
difference, but remember that disasters begin and end locally.
Changes need to happen at all levels of government.

Call your representatives. Your representatives' phone numbers
should be in your favorites list. Call your senators and representatives but don't forget to reach out to your local representatives too. Local politicians may be easier to get on the phone.
Sure, a lot of politicians are in the pockets of big oil and are
doing backroom deals with commercial real estate developers,
but there are some who will agree with you that the safety of
the community should be a priority. Even if they don't like what

you have to say, it's still their job to listen to their constituents' concerns. So don't back down.

You may find that they're not even aware of the problem you're working on, especially local politicians and those that are new to the job. If it has not been a big issue in your community before or covered by the media, it may not even be on their radar. In this instance, what you say to them can be particularly important. If you're the first person they've heard from about the problem, you have a huge opportunity to educate them and shape their thinking.

Review your local budget. Spend some time on your local government's website. Look for your town, city, or county budget and check to see how much is earmarked for emergency management. Unless you live in a big city, you are probably going to be surprised by how small the budget for your emergency management agency is. If you're in a really small town you might not even see emergency management in the budget.

Some questions that can help you assess if emergency management funding aligns with your community's needs: Are there any other sources of funding (like a state or federal grant) that is not represented clearly in the budget? How does the emergency management budget compare to other local agencies? Does this seem reasonable given the extent of risk your community faces? Have they put in a budget request for extra staffing positions or equipment in the recent past and been denied? What are the budgets of other nearby towns?

Vote. Voting is one of the most important things you can do if you want government to pay more attention to your community's needs. Voting locally has huge implications for emergency management. Your mayor has a big role in local emergency management; your state legislature can fund extra mitigation grants; your governor is the one that asks for a federal declaration during a disaster. At the local level, you have the opportunity not just to vote for the candidate who supports emergency management, but also to encourage them to include emergency management in their agenda. Especially if you live in a state that does not have constant disasters, emergency management may not be something these candidates have thought much about, and you can help educate them about local concerns.

Spread the word. Media coverage of emergency management issues is really important. You have to figure out the narrative, or story, that you're going to use to help people understand why the problem you're working to address matters. Why is this a problem everyone should care about when there are a million other issues that are demanding their attention? You will need a compelling story to persuade others into action and change. Look to see if there has been previous coverage of the problem you're working on. If there has been coverage, email the author and introduce yourself. They may want to talk to you for future stories. Speaking with journalists can be intimidating, but keep in mind that media coverage is an important and effective way that more people learn about both the problem and the solution.

You can submit an op-ed for consideration or write a letter to the editor. If you don't have any luck with traditional media, that's okay—move on to the internet. Write a blog post or social media post and share it with others who are working on similar issues. You may be surprised at how quickly this can gain traction as your experience resonates with others.

Emergency management isn't solely done by emergency managers—it involves all of us. We need your help in whatever way your resources and talents allow. It is okay if you don't have the time to dedicate your whole life to getting a flood wall fixed. Any hours you are able to contribute are important, even if it doesn't feel like much. Further, while it is important you work to educate yourself, you don't need to go out and get a master's degree. Anyone can do this work and we need *everyone* to do this work.

If you are a teacher, teach your students about climate change. If you are a writer, write about disaster recovery. If you are a cook, bring food to the community meeting. If you are an elected official, align your policy platform to incorporate emergency management reform. Take whatever corner of the world you occupy and use your time and talents to protect it.[521]

We are now living in the future scientists have warned us of my entire life. We did not stop climate change and now we have to figure out how to live in our new climate. We need to make a plan for how we will manage the consequences. There is still much we can do to minimize the impact, but doing so requires us to simultaneously build the capacity of the emergency management system while dismantling the inequality within it, all while our risk increases. This won't be easy. Controlling the damage will take all of our brilliant minds, passionate souls, and hard workers.

We need to aggressively hold accountable the people and organizations who are in positions to shape these conversations. If you do not like the way mitigation or recovery is unfolding in your community, you have to be the one to change it. Like my mom and the activists in Taylorville realized, there is no one coming to save you. We have to save ourselves.

Climate change requires global action, but much of climate adaptation must happen locally. Local knowledge of needs and solutions is essential, and often the authority to mitigate our risks resides at the local level. We need to leverage resources—more than what local communities can come up with on their own. Doing so requires a national movement.

What would happen if we intentionally created an emergency management system that centers justice, prioritizes our actual risks, and is driven by empirical research? The solutions presented throughout this book are an immediate and necessary starting place for creating a more effective and just emergency management system. Although these policies are largely Band-Aids for the existing system, they do lay the groundwork on which radical and transformative community-driven policy can spring. But true disaster justice requires a complete upending of the emergency management system as we know it.

What does an approach to emergency management look like if the only rule is to meet the needs of people and communities? What if FEMA's goal was holistic recovery? What if the goal was not only to respond when disasters happened, but to completely prevent disasters altogether? What if we rebuilt our world to live with hazards, rather than in opposition or ignorance to them? What if our approach centered environmental justice by dismantling inequalities, rather than reinforcing them? At the dawn of the climate crisis, there has never been a greater need or better moment to do this kind of vision building.

None of these problems are easy to fix. It is messy, difficult, and frustrating. It involves radical social, political, and economic change. It is going to require an organized, concerted effort under the pressure of a ticking clock. Every minute we do not make changes, another home is destroyed, another community is lost, and entire cultures are threatened.

On Finding Courage

I WAS EXCITED to learn Fargo would host a climate march in April 2017. To my knowledge, Fargo had never hosted a climate march like this, so it felt like a positive step forward. Without people and places like Fargo involved, the possibility of climate action in time to stem the worst of its consequences is not likely.

Since the Women's March that January, there had been an increase in marches and rallies around town. We had marched through Fargo protesting the Muslim travel ban, done a sister protest for Standing Rock, and confronted Representative Kevin Cramer at a town hall meeting. I had noticed this was all very new for many of the people attending and organizing these events. Although they were passionate, I had observed a timidity that was unusual compared to the type of people I had encountered at protests in other parts of the country.

The lead organizer of the climate march asked if I would be willing to speak. He had been struggling to find climate experts in North Dakota who were willing to risk the possible political repercussions of speaking at the event. I had spoken at protests in the past, but I hadn't spoken at one since I had been in graduate school. In many corners of academia there are people drawing lines that we are told not to cross.

I had been so immersed in academia for so many years that I had almost forgotten why I was there—to create change. So I followed the lead of scholar activists and wrote a speech that married the empirical disaster research with a call for policy change. I wanted to use this moment to show how the work I had done in academia could be used to create change.

On the day of the march, I looked out at the crowd and read signs that suggested we should care about the climate crisis because of the impact on the planet and for the benefit of our grandchildren. I had to smile because this was exactly the narrative I meant to correct in my speech. The consequences of the climate crisis were occurring as we spoke. This was not a battle to fight for our grandchildren, but rather for us.

I shared the many ways that climate change will affect us all in our lifetimes and, more urgently, listed the names of places around the world that were already feeling the impacts. Many lifelong climate activists would not have found anything new or novel in my speech, but as I spoke to people from the crowd afterward, it seemed like what I had said resonated. A few of them told me they had learned something. It was the first time I felt confident in how I could merge my disaster work and climate activism. I saw how my skill—explaining disasters—could impact the climate movement and that I could be one of the many who worked to address this need.

After the speeches, we marched around downtown Fargo and I felt, for a moment, hopeful about our future. Here, in one of the most conservative cities in America, a place that thrived off oil industry profits, the same state where Water Protectors had been met with state-sanctioned violence just months before, hundreds of people, mostly white suburbanites who seemed to have a newfound appreciation of the urgency of the climate crisis, gathered together.

We headed north from Island Park to the post office and then looped back south. We streamed past the little boutiques, cozy coffee shops, and college bars. Hundreds of North Dakotans marched confidently past the local news cameras set up intermittently along the route. The air filled with chants and there were a few honks from passing cars, presumably of support. It felt like forward movement. The possibility of a climate march like this one just five years earlier when I first moved to Fargo was unimaginable. I was grateful to be able to see this little pocket of the country begin to find their power.

As we neared the end of the march we were brought to an abrupt stop at the railroad tracks that run through the middle of town. The lights flashed and the gates closed as a train passed. Car after car of Bakken Oil flew by us from the oil fields of western North Dakota. The marchers' signs, with pleas for renewable energy, flapped in its breeze.

I looked around at my fellow marchers and found them seemingly unbothered. They stopped walking, tucked their signs under their arms, chatted among themselves, and scrolled through their phones, waiting patiently for the train to pass. I stood there stunned by their reaction and resisted the urge to scream, "HELLO! THERE IS A TRAIN FULL OF OIL INTERUPTING OUR CLIMATE MARCH!" I wondered if they did fully grasp what was at stake.

Hollywood always finds a way to give disaster movies a happy ending so perhaps it's not surprising that I've noticed there is an expectation that climate—and disaster—stories be suffused with hope. Even journalists manage to sprinkle a bit of hope through their disaster coverage. A favorite story newscasters offer during disasters is the one about a small miracle that has occurred amid the destruction. It happens so often that there are tropes:

an old man who survived days on his roof, the cat that wanders home weeks after a tornado, and the young child who is pulled from earthquake rubble still alive.

As I have gone from disaster to disaster, I've come to see that these stories aren't unique. You can find them in any disaster. The novelty of these stories starts to fade quickly when you see how frequently they happen. I don't mean to be cynical, it's just that once you understand why disasters happen you begin to see through these stories. They aren't miracles. The kid was found in the rubble because elite USAR (urban search and rescue) teams were flown in from around the world, and the guy was rescued from his roof because of volunteer search teams. Any feeling of hope I might have from the rescue is drowned out in the despair of yet another disaster. What about the hundreds of others who weren't found alive in the rubble? Where was their miracle?

These aren't hopeful stories to me; they are evidence of failure. Why was the man on his roof in the first place? Could he not afford to evacuate? Was he not given enough of a warning? Why did the building collapse on the child? Why were there not stricter building codes? Our communities shouldn't have to rely on miracles. These stories should not leave you hopeful; they should leave you furious.

Yet, every book I read about disasters or climate change seems to end on something meant to leave the reader feeling hopeful. For many years, I believed I too had to tell stories that have happy endings. As I started writing and speaking about disasters, I noticed I was asked one question again and again. What makes you hopeful about the future? This question spills out of my inbox in the emails I get from people around the world who are wanting some reassurance about their community's future. It's among the first questions I'm asked when I give a talk about climate. And, from my students after particularly tough lectures.

I used to search disasters for these "signs of hope" so I could appease these questions. I half-heartedly ascribed meaning to the single flower that grew from a crack in the sidewalk and wove metaphors about the cross of the church that remained standing in a sea of destruction. I told elaborate stories of how it is the people who drop everything to go volunteer and the survivors who put aside their differences to save each other that gave me hope.

I do not go looking for hope anymore. I do not go looking for the flower growing through the sidewalk crack. I know it'll be there, but I also know that flower is not going to get the neighborhood rebuilt more quickly. That flower is not going to stop the next disaster or save someone from their roof. I have seen the start of climate change's death toll and the nearly silent reaction to it. I have stood in too many destroyed communities and watched the system we have fail to meet survivors' needs too many times to feel hopeful about the climate crisis. Mostly, though, I realized that for me hope isn't a prerequisite for action.

If you need hope to stay motivated, that's okay. There are people you can go to for that. I will ask though that you do not become distracted by your search for hope or moved to inaction by its assurance once you find it. However, if you, like me, are looking for hope only because you see others doing it, consider this your permission to stop. The second I stopped apologizing for not feeling hopeful I found something better—courage.

Climate scientist Dr. Kate Marvel is the first person I heard talking about the importance of climate courage. In an essay on the subject she wrote, "Courage is the resolve to do well without the assurance of a happy ending."[522] There is no assurance that the climate crisis will have a happy ending. There are already too many disaster victims of climate-driven hazards who did not get a happy ending. There are already too many front-line communities for whom it is too late for a happy ending. But we do the work anyway.

It is courage that gets me out of bed each morning to go to

work and keep fighting for our future. My courage is driven by anger at seeing the injustice of others in pain. Courage gets me writing. Courage makes me speak at rallies and march in protests.

When I look to survivors, it's not hope that has led them to action, it's courage. The mothers of Taylorville had courage, just as New Orleanians did post-Katrina, and the communities of southeast Louisiana have now. Hope does not get flood protection systems built. It does not get homes bought out. It does not get infrastructure repaired. Some may also have hope but it is courage that propels disaster survivors and activists.

Being courageous requires asking hard questions that many in the United States have thus far had the privilege of not answering. What is an acceptable number of disasters? An acceptable number of lives destroyed? An acceptable economic impact? Who will pay for the damage? Who will decide when communities can no longer live where they have for generations? What is an acceptable number of deaths?

Being courageous requires us to make choices. We have to choose to prioritize the health of our ecosystems. We have to choose to minimize the suffering in our communities. We have to choose to question the capitalist, patriarchal, and racist systems that generate risk. We have to choose to build a more effective emergency management system. We have to choose to account for injustice as we manage our risk. We have to choose to prioritize our collective future. We have to choose to prioritize our safety. We have to choose to take responsibility. We have to choose integrity and justice.

Hurricanes get worse before they get better, but it is the promise that the storm will eventually end that gets us through it. We do not have that guarantee with the climate crisis. It won't just eventually pass. We have to intervene to end the crisis. Every minute of climate inaction and not preparing for the consequences keeps us in the storm with no end in sight.

I want to survive this storm.

I want us—all of us—to survive.

I am making a *choice* to fight for our survival and I hope you will too.

EPILOGUE

"But people take the news of their doom in diverse ways."

—SUSAN SONTAG

I WAS SUPPOSED to spend the summer of 2020 on a Great American Road Trip with my little sister. We were going to depart from Omaha, Nebraska, and drive west through Colorado and Nevada to Oregon. We were going to zigzag back and forth across California, ending in Joshua Tree before going to Las Vegas and the Hoover Dam. We were going to camp at the Grand Canyon and then spend a week in Santa Fe before driving across Texas to Houston and the coast. We were going to stop in New Orleans to see old friends and go up to Selma to see the Edmund Pettus Bridge and then drive back down to the Florida Panhandle. If we weren't sick of each other by then, we were going to loop the entire coast of Florida. We were negotiating the logistics of a detour to Puerto Rico before heading to Savannah, the Carolina islands, Norfolk, and a bunch of little towns in between. We were going to do a night each in DC and New York City to look at colleges, just a year away for her, before arriving at our family's house on the coast of Southern Maine.

The intention of this trip was to see for ourselves the impacts of the climate crisis across the country as a whole. We would have visited recently flooded farmland in Nebraska and seen the quickly melting glaciers in Colorado. I was going to meet with emergency managers on the West Coast who had been working wildfire responses nonstop and emergency managers in Arizona where heat-related deaths are rising. We were going to skip along the coastal states to check in with community organizers leading hurricane and flood recoveries.

We were going to drive down Louisiana Highway 1 to Grand Isle for a snowball and to see what has changed since I was last there. We were going to go to Vicksburg, Mississippi, to see the National Military Park that was used as an evacuation site during the 1927 Mississippi River flood. We were going to see the Florida Keys, where neither of us has ever been, and check on Miami clasping on to what is left of dry land. We were going to go to the Jersey Shore and then to Long Island, where our dad grew up, to see year eight of Sandy's recovery. Then we were going to wind up the coast through all the little New England neighborhoods that look like Camp Ellis.

We were going to bear witness to the pain and the damage climate change has already brought, but we were also going to meet the people who are fighting to fix it. We were going to talk to the people who are trying to ready us for a future that has already arrived. I planned to write about these places and these people, to elevate their stories and argue that we must support their work, especially in the absence of government-led action. Mostly, though, I wanted to show my sister, just sixteen, that there are good people in every corner of the country working hard for her future.

COVID thwarted our plans. The pandemic threatened to bring my work, and the work of others, on the climate crisis to a grinding halt. I kept thinking of my time in New Orleans post-Katrina when I did not yet understand the connection between climate change and disaster. A time when I thought recovering from one catastrophe meant we had the luxury to stop working to prevent future catastrophes. It was a time before I understood they are all connected. We can't stop talking about climate change every time there is a crisis, because then we would never talk about climate change.

The pandemic also reminds us that disasters do not wait their turn. The beginning of March 2020 was marked by a tornado destroying hundreds of homes in Tennessee and ended with twenty-two people injured by a tornado in Jonesboro, Arkansas. On a particularly windy night in April a quarter of a million Mainers lost power and a tornado outbreak Easter Sunday killed thirty-six across the Southeast. May began with another earthquake in Puerto Rico. Then two poorly maintained dams in Michigan were overwhelmed by significant rainfall. Ten thousand residents were put under evacuation orders as nearby towns including Midland, home of the Dow Chemical Company headquarters, were flooded. Wildfires were frequent across Florida and the West prompting evacuation orders. A derecho blew through the Midwest in mid-August causing an estimated $7.5 billion in damage.[523] The National Weather Service issued its first ever fire tornado warning in the same week "twin hurricanes" danced around the Gulf of Mexico before making landfall in Louisiana within days of each other. Southwest Louisiana was hit hard by Hurricanes Laura and Delta and the eye of Zeta went directly over the city of New Orleans. By the end of August record-breaking wildfires were again burning in California and smoke from the West threatened the rest of the country. By October wildfires in Colorado broke state records as even more people were forced to evacuate their homes.

Outside the United States, Cyclone Harold cut across Vanuatu and millions of people in India and Bangladesh had to evacuate for Cyclone Amphan. There was a landslide at a jade mine in Myanmar. By July millions of people across China were affected by flooding as the controversial Three Gorges Dam reached its capacity. Even Wuhan, the origin of COVID-19, was threatened by floodwaters. An explosion in Beirut killed 180 as response efforts were made more difficult by political turmoil, economic crisis, and the pandemic. Hurricane Eta was quickly followed by Iota, which both caused extensive damage across parts of Central America.

We did not get to ignore those crises because we were already in the middle of a pandemic. The consequences of climate change do not wait until it is convenient for us to manage them. A critical lesson of the pandemic—specifically the failed federal response—is that we are not prepared to manage the consequences of the changing climate. We have the knowledge of how to do better, but those in power have to actually choose to use it. There are no plans, no policies, no politicians, that will protect us unless we demand it.

On January 21, 2021, just two weeks after violent white supremacists led an insurrection at the Capitol, I watched President Joe Biden and Vice President Kamala Harris be sworn into office. Poet and activist Amanda Gorman recited the inaugural poem. In it she reminded us that we cannot allow catastrophe to prevail over us. Her words about our history, this moment, and our future echoed across Washington, the country, and the world. We have to prevail.

The Trouble with Climate Havens

SINCE THE INITIAL publication of *Disasterology* in 2021, the US has been through just under one hundred disasters, each having cost more than a billion dollars.[1] I have continued traveling around the country to see the devastation and hear the stories of survivors and emergency managers. One of the places I have returned to repeatedly is New Orleans. In the past few years, several storms have affected the coast of Louisiana, but for the southeastern part of the state, Hurricane Ida was most damaging.[2] Across the state, there have been wildfires[3] and wetland

1   NOAA. (2024). "Billion-Dollar Weather and Climate Disasters." National Centers for Environmental Information. https://www.ncei.noaa.gov/access/billions/#:~:text=In%202024%20(as%20of%20September,and%202%20winter%20storm%20events.

2   Reckdahl, K., Goodman, J. D., and Brasted, C. (2021, August 21). "Hurricane Ida, a Powerful Category 4 Storm, Batters Louisiana." *New York Times*. https://www.nytimes.com/2021/08/29/us/hurricane-ida-a-powerful-category-4-storm-batters-louisiana.html.

3   Cline, S. (2023, September 18). "In a state used to hurricanes and flooding, Louisiana is battling an unprecedented wildfire season." AP News. https://apnews.com/article/louisiana-wildfire-b9d8968c1ce98b009c3ce95fa08a8f40.

fires[4] whose smoke on one occasion combined with dense fog and caused multiple accidents involving 158 cars[5]. In 2022 two tornadoes followed similar destructive paths through Arabi.[6] A toxic fire at a Marathon Petroleum refinery in Garyville has left residents concerned for their health.[7] Then there was the intense rainfall for several days that overwhelmed New Orleans's drainage system, leaving the streets underwater. New Orleans and the surrounding communities continue to be on the front lines of the climate crisis but perhaps none more so than Isle de Jean Charles.

Managed Retreat

On my last trip to Louisiana, I drove down a new road built just off the highway outside Houma. It is a new neighborhood with a community center, freshly laid sidewalks, and identical homes. Each has a front porch, a long driveway, and a manicured green lawn. Other houses were finished but vacant, a few were mid-build, and some were still just an empty plot of land. Nearby there is a water-treatment plant and a facility that deals in equipment for the oil and gas industry, but otherwise,

4   Baurick, T. (2023, November 4). "Smoky New Orleans swamp fire elicits slower, lighter response than other wildfires." *The Times-Picayune*. https://www.nola.com/news/environment/new-orleans-swamp-fire-gets-slower-lighter-response/article_28896af4-7a89-11ee-8765-2bddf2184d5a.html.

5   McGill, K. and Cline, S. (2023, October 25). "Toll rises to 8 dead, 63 hurt from Louisiana interstate pileup blamed on dense fog, marsh fire smoke." AP News. https://apnews.com/article/superfog-louisiana-highway-crashes-marsh-304195491d1bfbc5ac929b1c6acd1d7f.

6   Shannon, C. (2022, December 16). "How similar was this week's New Orleans-area tornado path to March 2022? This map shows both." *The Times-Picayune*. https://www.nola.com/news/weather/new-orleans-area-tornadoes-nine-months-apart-similar-paths/article_348d91b4-7d54-11ed-8de2-cbbc919a3877.html.

7   Laughland, O., Sneath, S., and Craft, W. (2024, September 9). "The huge US toxic fire shrouded in secrecy: 'I taste oil in my mouth.'" *The Guardian*. https://www.theguardian.com/us-news/article/2024/sep/09/marathon-oil-fire-louisiana-cancer-alley.

there is room for the neighborhood to grow. If you were passing through, you might not even notice the street; it looks just like any other American subdivision.

But it is not like the others. This is The New Isle, a "climate resettlement" and the new home of Isle de Jean Charles residents, or as they have called themselves in the past, "America's first climate refugees" (and more recently "climate pioneers"). Before seeing the new neighborhood, I had generally thought the effort to move the community inland was the necessary solution given the circumstances. And yet, as I drove the length of the road, I found myself completely disoriented, unable to muster anything more than a noncommittal statement of "Huh."

The residents of The New Isle come by way of a now-tiny island next to the Gulf of Mexico—Isle de Jean Charles. The island was settled in the 1800s by Indigenous people who had been pushed out of other parts of Louisiana by European colonizers.[8] When they arrived, Isle de Jean Charles was a sprawling thirty-four square miles. It took time to learn the new land, and the island was accessible only by boat, but the seafood was abundant. They began farming rice and corn and raising cattle. In the mid-1900s the state started providing public education to island residents and built a road connecting them to Pointe-aux-Chenes on the mainland. Greater government involvement also came with basic utilities including running water, electricity, and mail delivery. Isle de Jean Charles grew into a thriving coastal Louisiana community.

Among the first signs that Isle de Jean Charles would face an existential threat without significant intervention was Hurricane Carmen in 1974. The storm damage was severe enough that an initial wave of people left the island for the mainland. Another fifty families left in 2002 after Hurricane Lili. Erosion

8   McGrew, W. and Olivier, J. (2023, November 27). "In the New Isle." *Country Roads Magazine*. https://countryroadsmagazine.com/art-and-culture/people-places/new-isle-isle-de-jean-charles/.

caused by storm surges, man-made flood control, sea level rise, and the oil and gas industry extraction approaches took the land from under them. Today only half of a square mile is left. In only seventy years, 33.5 square miles have been lost.

Isle de Jean Charles is still accessible on days when the road to the island is above water. Along the road, you can see abandoned mailboxes peeking out of the overgrown brush marking where houses, now gone, once stood. The remaining homes are almost entirely built up on the high stilts common of coastal Louisiana architecture and are in various states of disrepair. Most have been pieced back together time and again after being battered by storms. As such, no two homes on the island are the same. Now that almost all residents have moved to The New Isle, the legal requirement to let the homes on the island decay as they weather is well underway.

It was not necessarily inevitable that the residents of Isle de Jean Charles would need to abandon their homes. In 2002, the Corps of Engineers announced plans for Morganza to the Gulf, a flood-protection system for southeastern Louisiana. It included ninety-eight miles of new levees built to protect against a Category 3 storm and was projected to cost $3 billion. Isle de Jean Charles was excluded from the plan. This was a death sentence for the island, not only because it would be outside the protection of the levee system, but also because studies showed the new levees would actually increase flooding on the island. The Corps of Engineers justified their decision based on the financial value of Isle de Jean Charles. In the eyes of the federal government, the community was not worth the money it would cost to protect it from flooding.

Given the rapid land loss, the residents immediately began developing plans to move. The state was supportive of their efforts to secure federal funding, which they would need. Without funding, many residents would be unable to afford to leave the island. They also wanted to find a way to stay together to preserve their culture. This would require more than the government's buying out individual homeowners; rather, a full community relocation would be needed. Together they would build an entirely new neighborhood somewhere else.

Not everyone on the island thought this was necessary, but Hurricane Isaac's damage in 2012 was significant enough that most remaining residents accepted the severity of the situation. Planning efforts escalated, and in 2014 Louisiana won a $48.3 million competitive resilience grant from the Department of Housing and Urban Development to build the new neighborhood.[9] For the next decade, nonprofits along with federal, state, local, and tribal governments negotiated the details of the move. They had to find answers to numerous questions: Which residents were eligible to receive a new house? How much would they be compensated? Where would the new neighborhood be? What would the houses look like? Who would build them? What would happen to the houses left behind? How would they preserve their culture, which is intertwined with the coast?

This was a brutal process.[10] It was often unfriendly and often not handled with care. There were divisions between the various organizations and agencies involved and internal divisions among residents. This process did not unfold in a straight line, and at times it was not approached in good faith. So on the drive from New Orleans to The New Isle, I was excited to see the product of these efforts. There had been many times in the past decade when it did not seem likely that the relocation would actually happen. If it failed, it would not just be a problem for the resi-

---

9   Loginova, O. and Cassel, Z. (2022, August 17). "Leaving the island: The messy, contentious reality of climate relocation." *The Center for Public Integrity*. https://publicintegrity.org/environment/harms-way/leaving-isle-de-jean-charles-climate-relocation/.

10   Dermansky, J. (2022, September 8). "Isle de Jean Charles Community Members Moved into the First Federally Funded Resettlement Project in Louisiana Despite Visible Engineering Issues." *DeSmog*. https://www.desmog.com/2022/09/08/isle-de-jean-charles-relocation-new-isle-climate-change/.

dents of Isle de Jean Charles—it could result in consequences for other communities in similar situations. I thought of Camp Ellis back home in Maine. The New Isle was meant to be the model. This is what the outcome of the climate movement's calls for relocating communities—even entire cities—away from the coastline would look like. So, although the road there had been bumpy, the fact that the neighborhood had been built meant, I thought, that I was about to see as happy an ending as possible—a way for us to manage climate risk more permanently.

Instead, as I drove through The New Isle, my heart sank. I was surprised to be struck by an overwhelming feeling of grief. Perhaps it is my tendency toward pessimism, but as I looked around, all I could think about was what had been lost. After all, the "solutions" to adapting to climate change are rooted in the consequences of climate change. I felt like I was mourning instead of celebrating. The New Isle is so *different* from Isle de Jean Charles. I hope for the people who live there that The New Isle will be a safe place where they can heal and build a thriving future for themselves. But this also is not something I can imagine wanting for others. We must work really hard to prevent communities from finding themselves with no other option but to move, and develop a better approach to moving communities when necessary.

I have always been wary of the thinking that managed retreat is the primary tool of climate adaptation, as it is often presented. To me, physically moving has always been one mitigation approach that can be used in the most dire of circumstances. The argument for doing as much hazard mitigation as possible, before picking up and leaving, is strong. Isle de Jean Charles did not start managing the erosion problem by deciding to move inland. They started by taking other preventative measures that allowed them to live on their land as long as possible—rebuilding the road and raising their houses. It was only once the government announced they would stop funding mitigation efforts—in the exclusion from Morganza, and the threat of denial of services, and as the land literally disappeared—that they ran out of options.

The logistics and costs required for managed retreats (especially of entire cities) and the turmoil they cause are extensive. I think of how hard the people of Isle de Jean Charles worked to be able to move. They had time, money, the mobilization of local, state, and federal governments, a sympathetic story, and local media that supported their efforts. They, like Camp Ellis, had a lot working in their favor to ensure a successful resettlement. But as of today, The New Isle is home to twelve houses. Is twenty years and millions of dollars for a dozen homes really a model for other communities? I hope not.

Underlying climate migration is *who* is told no other options exist. The language of the climate crisis—including terms like *resettlement*, *relocation*, *refugee*, and *pioneer*—is rooted in the language of colonialism. Many people—particularly marginalized communities—will not independently decide to move on their own and have the resources to do so. Instead, their hand will be forced by government decisions, like when the government said they would stop maintaining the road to the island, or in the post-Katrina policies that made it impossible for some to return to the Lower Ninth Ward.

Some argue it is better for the government to be involved, to take the approach of "managed" retreat, because if moving is left unmanaged by government, people will likely lose their life savings and the process could be even more inequitable. For example, if government does not intervene with the insurance markets, we will see even more people become trapped in dangerous locations without insurance as a safety net. They will struggle to sell their houses, because who wants to buy an uninsurable house? They risk losing everything.

Isle de Jean Charles is not completely abandoned. There is still an active marina and people fishing alongside the road to the island when it is not submerged. Louisiana just spent millions building a riprap seawall on one side of the road despite saying they would no longer invest in infrastructure for the island. It is not unusual for the government to turn the land where buyouts have occurred into public recreation areas like parks. It is much easier and cheaper to rebuild a park after a flood than dozens of houses. Yet there are persistent concerns that developers may be allowed to build new fishing cabins on the island.[11] In this way, worries that Isle de Jean Charles may turn out to be another story of climate gentrification remain.

The purpose of managed retreat is to move people out of harm's way to safer places. The New Isle is a fifty-mile drive away from Isle de Jean Charles. But the topography of southeast Louisiana is such that storm impacts, including flooding, can be felt far inland. While The New Isle certainly has less risk than the island, it can still flood. In 2021 as Hurricane Ida made landfall, its effects were felt not just on the island but also in The New Isle. Some residents were delayed in moving because the storms had caused damage. One house, on the day of closing, had standing water underneath it.[12] So, while it may be the case that these new locations are safer than the places left behind, we need to understand that there is no place we can go to eliminate risk entirely.

Climate Havens

I had long wanted to visit Asheville, North Carolina, the liberal, artsy hub of Appalachia. In August 2024 I made the trip. I

11   The Courier and Daily Comet. (2021, August 22). "Company drops plans for subdivision on Isle de Jean Charles." *Houma Today*. https://www.houmatoday.com/story/news/2021/08/22/company-drops-plans-subdivision-isle-de-jean-charles/8229713002/.

12   Dermansky, J. (2022, September 8). "Isle de Jean Charles Community Members Moved into the First Federally Funded Resettlement Project in Louisiana Despite Visible Engineering Issues." *DeSmog*. https://www.desmog.com/2022/09/08/isle-de-jean-charles-relocation-new-isle-climate-change/.

stayed at a house precariously perched on a steep slope just outside the city. I wandered around downtown and popped into the little shops and art galleries. I strolled through the weekend farmers market and visited the North Carolina Arboretum. I went to a weird little cat film festival at a theater in the art district and went tubing down the French Broad River.

Weeks later that same river overflowed its banks, flooded the movie theater, and much more. The trees at the arboretum came crashing down, and the shops in town were closed. I wondered if the house was still there. I watched Hurricane Helene from afar and learned the true extent of the damage as cell service began to return after many days. The geographic scope of Helene's path was extensive—nearly five hundred miles from Florida's Big Bend to Southern Appalachia.[13] The full economic cost is not yet known at the time of this writing, but it will be among the most expensive hurricanes in US history. The total death toll is also unknown, but it is at least two hundred people, making Helene the third-deadliest US hurricane of the twenty-first century behind Hurricanes Maria and Katrina.[14]

As the rain began to fall, the emergency-management system—both formal and informal—swung into gear. Despite the widespread mis- and disinformation spread by far-right Republicans, FEMA's response was surprisingly effective given the circumstances.[15] The emergent response from the affected communities was as strong as they come. While the coming

---

13 Ramirez, R., Paget, S., Fisher, A., and Merrill, C. (2024, October 1). "What Hurricane Helene's 500-mile path of destruction looks like." CNN. https://www.cnn.com/2024/10/01/weather/hurricane-helene-path-of-destruction-climate-dg/index.html.

14 Jervis, R. (2024, October 7). "From rescue to recovery: The grim task in flood-ravaged western North Carolina." *USA Today.* https://www.usatoday.com/story/news/nation/2024/10/07/helene-flooding-rescue-recovery-north-carolina/75545453007/.

15 Ahmed, A. (2024, October 17). "The Hurricane Conspiracies Made It Clear—We're Going Climate Delulu." *Atmos.* https://atmos.earth/the-hurricane-conspiracies-made-it-clear-were-going-climate-delulu/.

together of survivors is a hallmark of disaster response, Asheville's exceptionally strong community networks enabled a rapid homegrown response. Even so, the damage is expansive. Recovery will take decades.

For many, the devastation in Western North Carolina was a shock. That a hurricane could cause such destruction so far inland once again rattled the misperception of climate change as a coastal problem. It left people wondering how the so-called climate haven of Asheville could be devastated by a climate-fueled hurricane before the frontline climate city of Miami.

Unlike New Orleans and Isle de Jean Charles, which have long been on the front lines of the climate crisis, other cities have been recently touted as climate havens. Few in the academic literature use this term; instead, it mostly comes up online in lists that identify cities in the United States deemed safe from the effects of climate change. Asheville made regular appearances on these lists besides places like Buffalo, New York; Burlington, Vermont; and Duluth, Minnesota. These lists often come with a caveat that nowhere is completely immune from climate change, but by the time you get that far into the article, it is too late. The lure of the term *climate haven* is strong. Being told there is a place you can go to avoid disaster is seductive.

Perhaps most worrying, some cities marketed themselves as climate havens to attract new residents and grow their economies. These campaigns made an implicit promise—if you come here, you will be safe. Who wouldn't want that? These campaigns seemed to work. Op-eds were written by people leaving California for Vermont to get away from the fires.[16] In Asheville, there was a local news story featuring a real estate agent who said his clients were moving to the city specifically because it was

16   Leslie, J. (2021, May 23). "Op-Ed: Leaving California." *Los Angeles Times*. https://www.latimes.com/opinion/story/2021-05-23/leaving-california-climate-change-wildfire.

a climate haven.[17] The problem is Vermont and North Carolina are not climate havens because climate havens do not exist.

In the days following Helene, *Vox* tracked down the origins of the *climate haven* term.[18] A real estate researcher accidentally popularized the term after alluding to it in an interview about where people could move to get away from climate impacts. The catchy "climate-proof Duluth" slogan suggested in an economic-development package for the city began as a joke.

Buffalo was another city that latched on to the term.[19] Climate change, with its warming temperatures, the city argued, would minimize the brutality of Buffalo winters. Plus, the city isn't on the coast. While it is true Buffalo faces fewer and less intense hazards generally than some other parts of the country, that is a far stretch from being a "haven." Despite not being on a coastline, Buffalo still has a meaningful chance of flooding, both from intense rainfall and from the lakeshore. There is a long list of hazards from wildfires and heat waves to cyberattacks and dam failures that could happen. The range of hazards that could affect Buffalo are not even what made it so ludicrous to call the city a climate haven, though.

At the time, Buffalo, a city of a quarter million people, did not have a full-time emergency manager. On paper the person responsible was the fire commissioner, but very little was being done. For the Erie County Hazard Mitigation Plan published March 2022, the city reported they had no comprehensive emer-

---

17   Simon, J. (2024, October 9). "They came to Asheville looking for a 'climate haven.' Then came Hurricane Helene." NPR. https://www.npr.org/2024/10/09/nx-s1-5137024/climate-haven-hurricane-helene-asheville.

18   Estes, A. C. (2024, October 9). "The shady origins of the climate haven myth." *Vox*. https://www.vox.com/climate/377199/hurricane-milton-helene-climate-haven-risk.

19   De Socio, M. (2024, July 1). "US cities are advertising themselves as 'climate havens.' But can they actually protect residents from extreme weather?" BBC. https://www.bbc.com/future/article/20240628-us-climate-havens-cities-claim-extreme-weather-protection.

gency-management plan, no risk assessment, no recovery plan, no public health plan, and no continuity-of-operations plan for the city.[20] They also did not have a mitigation planning committee, nor did they have basic provisions identified like emergency shelters or temporary-housing locations.

No place is a climate haven, but especially not places that are not doing emergency management. The function of emergency management is to have someone who is thinking about and addressing risk across mitigation, preparedness, response, and recovery. Surely this is a precursor to claiming safety. It is the bare minimum.

By December 2022, tragically, the irresponsibility of Buffalo's city government in not hiring an emergency manager was laid bare when a major snowstorm caught the city flat-footed. Over forty people were killed, a notably high death toll for a winter storm in the northeast these days.[21] Those are death toll numbers we would expect to see with a winter storm in the south where there is typically a much lower level of preparedness for winter weather. A number of factors led to these deaths, but one major finding of post-storm investigations was the lack of coordination among city and state officials and agencies.[22] Considering it is the job of emergency management to coordinate city and state officials and agencies, it is not a leap to say that some, or perhaps many, of those deaths, could have been pre-

20   Erie County. (2022). Hazard Mitigation Plan, Volume II. https://www3.erie.gov/dhses/sites/www3.erie.gov.dhses/files/2022-04/hazmit_2022_vol2_0.pdf.

21   Watson, S.T. (2023, January 19). "Death toll in Buffalo blizzard rises to 47 people." *The Buffalo News*. https://buffalonews.com/news/local/death-toll-in-buffalo-blizzard-rises-to-47-people/article_04c578e0-9814-11ed-b391-dbf7d2370f3d.html.

22   Kaufman, S. M., et al. (2023). *Lessons Learned from the Buffalo Blizzard*. Robert F. Wagner School of Public Service. New York University. https://wagner.nyu.edu/files/faculty/publications/NYU%20Buffalo%20Blizzard%20Report%20-%20June2023_0.pdf.

vented if the city had invested in establishing an emergency-management agency.

Yet even in the aftermath of the storm, it still was not immediately obvious to some in local government that a city of a quarter million people urgently needed its own emergency manager. Local politics dragged on with some members of the city council threatening to not approve any other new hires for the city until an emergency manager was hired.[23] Eventually, in response to the failures of the winter storm, the city hired a fleet manager for snowplows and other heavy equipment, and also an emergency-services director who is meant, as best as I can tell, to fulfill the role of emergency management for the city. Even so, Buffalo is a big city. They need a robust functional agency with multiple full-time staff members who are able to dedicate their time across mitigation, preparedness, response, and recovery efforts.

The term *climate haven* is an invitation to move someplace safe. But if the underlying assumption of safety is wrong, it means we have people living in places where they do not understand the nuances of the risks they face. This takes away people's agency and can be used as an excuse to not invest in emergency management. I understand the desire to not think about disasters and climate change all the time, but not thinking about it, turning off our disaster vision, especially for elected officials, can be deadly. Unfortunately, a consequence of the delayed action to address the climate crisis means the luxury of not thinking about climate-related disasters is long behind us. Obscuring risk in any form makes us less safe.

There are people who moved to Buffalo, Vermont, and Asheville because they wanted to make a responsible choice for their

23 Kaufman, S. M., et al. (2023). *Lessons Learned from the Buffalo Blizzard*. Robert F. Wagner School of Public Service. New York University. https://wagner.nyu.edu/files/faculty/publications/NYU%20Buffalo%20Blizzard%20Report%20-%20June2023_0.pdf.

future. I do not blame them for doing so. We tend to make the best disaster-related decisions we can with the resources and information we have available to us. If you have the ability to factor in risk when you decide where you want to live, you should. I do not recommend buying a newly built house right on the ocean or moving into a new neighborhood in the wildland-urban interface that only has one road in and out. We should make smart decisions about risk when we can, but ultimately we are all making bets on where it is safest for us to live. We are deciding which hazards we are most willing and able to put up with (I have a personal preference of flooding over fire, for example) and who we want to have around us when disaster strikes.

Disaster researchers and emergency managers have not done a good enough job presenting the risk of climate change—and disaster—to the public. We need to be better about explaining the complexities of risk. It is not just will this or that hazard happen here but how those hazards will interact with the vulnerabilities of our community, and what the likelihood is of these interactions happening. What is being misunderstood is that the safety and preparedness of a community is not just based on the various hazards they face, but also their vulnerability to those hazards. Outside of mitigating climate change itself, there is not much we can do to change the characteristics of the hazards themselves, but there is a lot we can do to reduce our vulnerabilities. In doing so, we can significantly minimize the impacts and needs experienced. There are no climate havens, but I wonder if there could be.

When I think about the kinds of communities that could fare best in the climate crisis, it is not necessarily the places on the climate-havens list. Instead I imagine communities whose states have made plentiful recovery resources available and invested heavily in emergency-management systems. They are places where the government is working to address the insurance crisis, not for the benefit of insurance companies, but for their residents. They are places with strong social networks where big investments have been made in safe housing, robust education, and access to healthcare. They are places with disaster experience and appreciate the complexities of the emergency-management system. They are places that have developed disaster subcultures that are rooted in experience and care for their communities. They are places where risk is well understood and confronted openly, not hidden. By these measures, nearly any place *could* be a climate haven.

Some people may have no choice except to move, but far more often climate change will mean learning to reduce our vulnerabilities so that when the fire or storm or heat wave starts, we can absorb it. We have to relearn how to live with hazards. That is the point of this book—we cannot outrun all of our risk, but there is a lot we can do to minimize the damage and prepare ourselves to go through response and recovery in effective and equitable ways.

There are often reports from the UN or other groups that estimate potentially hundreds of millions of people being displaced by climate-related disasters—some who will need to permanently move, including entire nations.[24] And all I can think of are those dozen little houses in The New Isle. The world we used to live in is gone. When real estate marketers and mayors describe climate havens, they are trying to evoke the feeling of the "old world," but these climate havens are not immune to disasters—they just haven't been hit yet. Climate change has created a new world, and it is our job to figure out how to live in it. We need to transform all of our communities into safe places to live—we have to rebuild every single one of our communities to survive our new climate—which is what is meant when

---

24 Watson, J. (2022, July 28). "Climate change is already fueling global migration. The world isn't ready to meet people's changing needs, experts say." PBS. https://www.pbs.org/newshour/world/climate-change-is-already-fueling-global-migration-the-world-isnt-ready-to-meet-peoples-needs-experts-say.

we say we must *adapt* to climate change. And emergency management is at the center of helping our communities do that.

In the years since *Disasterology* first came out, I have become even more certain that the public—and policymakers—does not understand the scope, scale, and urgency of our need to adapt to the climate crisis. With the end of the Biden administration, although they regularly verbalized their support for climate action, their policies failed to meaningfully move the needle enough to meet the scale of the adaptation crisis. There have been minor policy changes within FEMA, particularly related to individual and household recovery programs, but the effect is not yet known, and even if they are positive changes, they will still fall far short of what is needed.[25] We have not spent nearly enough time, money, or resources on solving the problem of climate adaptation. There continues to be no appetite among national politicians to take it on.

We are going to see bigger, deadlier, and more frequent disasters in increasingly unexpected places. That is the "change" part of climate change. Appalachia and southern Louisiana share the distinction of being places where the causes and consequences of the climate crisis intermingle. I expect they too will be the places that create the path for adaptation, if only because they have no other option.

★ ★ ★ ★ ★

25   Flavelle, C. (2024, January 19). "As Climate Shocks Worsen, U.S. Disaster Agency Tries a New Approach to Aid." *New York Times*. https://www.nytimes.com/2024/01/19/climate/fema-disaster-aid-climate.html.

ACKNOWLEDGMENTS

THE MARK OF many can be found within the pages of this book. There have been people at every turn who have pointed—or pushed—me in the right direction.

Thank you to Greg Szkarlat who first took me to New Orleans. Thank you to my professors at Loyola University New Orleans, who were the first to teach me about disasters. Dr. Jessica Jensen, Dr. Carol Cwiak, Dr. Sarah Kirkpatrick, and Dr. Daniel Klenow at North Dakota State University have surely had the most significant impact on my understanding of emergency management. I am forever indebted to their generosity and brilliance.

I am particularly grateful to Dr. Amanda Savitt who has guided me forward. Her mark on this book is pronounced—not only in the placement of commas, but in the hours and hours spent talking through every concept, idea, and story. Sometimes multiple calls a day! This book simply would not exist without her.

Thank you to my dear friends Charles Bourg and Hayden Golden who both provided feedback on drafts. Thank you to John Carr for being an exceptional hype man and consulting on all pop culture references mentioned in this book. Thank you also to Bethany O'Meara and Dr. David Zelaya who have always provided me inexhaustible support.

Thank you to my book agent, Tess Callero, who found me at the exact right moment. She saw what this book could be immediately and was patient with me until we found the right home for it. Thank you also to my editor Laura Brown and the entire team at Park Row Books.

Thank you to my family—Peter, Karen, Parker, Jason, and Sara, who listened to me agonize over this book for years, supported me while I wrote it, and let me share our family stories. I am especially grateful for my sister Sara who read every word, told me when I sounded old, and edited out a few terrible jokes. Most of all I'm thankful to my mother, Karen, who read every draft and provided immeasurable support.

Thank you also to the survivors and people who have shared their stories and trusted me with them. I hope this helps.

This book is the product not only of people but places too. I wrote this book all over the country in New Orleans, Houston, Fargo, Omaha, Portland, and lots of places in between. I wrote it on planes and trains, alongside the ocean, rivers, lakes, and pools, at small coffee shops and big city libraries. Each of these places gave me a future to fight for and a world to love.

Endnotes

Introduction

1 For more on the crisis in Taylorville see: Menderski, M. (2014, September 13). "Superfund legacy remains in Taylorville and beyond." *The State Journal-Register*. https://www.sj-r.com/article/20140913/NEWS/140919731.

2 For more on the science of climate change see: IPCC (2014). "Climate Change 2014: Synthesis Report." https://www.ipcc.ch/report/ar5/syr.

3 Thunberg, G. (2018, November). "School strike for climate—save the world by changing the rules." [Video]. TED Conferences. https://www.ted.com/talks/greta_thunberg_school_strike_for_climate_save_the_world_by_changing_the_rules/transcript?language=en&fbclid=IwAR30UAigKwxy FfepGjlo7zaKWdWg-llQsd7vQQCJaJ_uUbStNI1IXWjaZnc.

For more on Greta Thunberg's climate activism see: Thunberg, G. (2019). *No One Is Too Small to Make a Difference: Illustrated Edition*. Penguin UK.

4 The latest global emissions gap report: "United Nations Environment Programme (2019)." *Emissions Gap Report 2019*. UNEP, Nairobi. https://wedocs.unep.org/bitstream/handle/20.500.11822/30797/EGR2019.pdf?sequence=1&isAllowed=y.

5 NOAA. (2020). "A Decade Later: Advances in Oil Spill Science Since Deepwater Horizon." https://response.restoration.noaa.gov/decade-later-advances-oil-spill-science-deepwater-horizon.

For more on the BP explosion read: Achenbach, J. (2012). *A Hole at the Bottom of the Sea: The Race to Kill the BP Oil Gusher*. Simon & Schuster.

For more on the causes and impact of the BP disaster read: Juhasz, Antonia. (2011). *Black Tide: The Devastating Impact of the Gulf Oil Spill*. John Wiley and Sons.

For more on the history of BP's corporate safety record read: Steffy, L. C. (2011). *Drowning in Oil: BP & the Reckless Pursuit of Profit*. McGraw-Hill.

6   For more on South Louisiana read: Tidwell, M. (2007). *Bayou Farewell: The Rich Life and Tragic Death of Louisiana's Cajun Coast*. Vintage.

7   Hall, S. (2015, October 26). "Exxon Knew about Climate Change almost 40 years ago." *Scientific American*. https://www.scientificamerican.com/article/exxon-knew-about-climate-change-almost-40-years-ago.

8   For more on the individualized approach to climate change: Maniates, M. F. (2001). "Individualization: Plant a Tree, Buy a Bike, Save the World?" *Global Environmental Politics* 1(3), 31-52.

9   Federal Emergency Management Agency. (2007). Principles of Emergency Management Supplement.

10   Emergency Management Institute (2015). "Statement of the Emergency Management Doctoral Degree Holder/Seeker Focus Group." https://training.fema.gov/hiedu/docs/emgt%20doctoral%20degree%20holder.seeker%20points%20of%20consensus.pdf.

11   For more on the discipline of emergency management see: Jensen, J. (2011). "The argument for a disciplinary approach to emergency management higher education." In J. Hubbard, *Challenges of Emergency Management in Higher Education*, (18-47). Fairfax, VA: Public Entity Risk Institute.

12   For a guide to the Paris Agreement: Denchak, M. (2018, December 12). "Paris Climate Agreement: Everything You Need to Know." National Resources Defense Council. https://www.nrdc.org/stories/paris-climate-agreement-everything-you-need-know.

13   For a criticism of the Paris Agreement see: Gustin, G. (2019, November 7). "The Paris Climate Problem: A Dangerous Lack of Urgency." *Inside Climate News*. https://insideclimatenews.org/news/07112019/paris-climate-agreement-pledges-lack-urgency-ipcc-timeline-warning.

14   For more on the issue of accountability see: Selin, H. and Najam, A. (2015, December 14). "Paris Agreement on climate change: the good, the bad, and the ugly." *The Conversation*. https://theconversation.com/paris-agreement-on-climate-change-the-good-the-bad-and-the-ugly-52242.

15   Favreau, J. (2017, June 1). "Pod Save America." [Audio podcast]. https://crooked.com/podcast/more-fucked-than-we-think.

16   For more on current climate impacts see the National Climate Assessment: USGCRP, 2018. "Impacts, Risks, and Adaptation in the United States." *Fourth National Climate Assessment*, Volume II. US Global Change Research Program, Washington, DC. https://nca2018.globalchange.gov.

17   Cal Fire. (2020, November 3). "Top 20 Deadliest California Wildfires." https://moien.lu/wp-content/uploads/2020/08/top20_cal-fires-deidlech.pdf.

18   Thompson, A. (2020, December 22). "A Running List of Record-Breaking Natural Disasters in 2020." *Scientific American*. https://www.scientific-

american.com/article/a-running-list-of-record-breaking-natural-disasters-in-2020.

19   Samenow, J. (2017, September 22). "60 inches of rain fell from Hurricane Harvey in Texas, shattering U.S. storm record." *Washington Post*. https://www.washingtonpost.com/news/capital-weather-gang/wp/2017/08/29/harvey-marks-the-most-extreme-rain-event-in-u-s-history.

20   Laizer, B. (2020, November 9). "Louisiana included in 'cone of uncertainty' for 8th time during hurricane season 2020." WGNO. https://wgno.com/weather/tracking-the-tropics/louisiana-included-in-cone-of-uncertainty-for-8th-time-during-hurricane-season-2020.

21   Smith, A. B. (2018, January 8). "2017 U.S. billion-dollar weather and climate disasters: a historic year in context." Climate.gov. https://www.climate.gov/news-features/blogs/beyond-data/2017-us-billion-dollar-weather-and-climate-disasters-historic-year.

22   Frank, T. (2021, February 18). "'It's serious': FEMA's disaster fund will be broke by April." E&E News. https://www.eenews.net/climatewire/stories/1063725355.

PART ONE

Recovery: The Second Disaster

23   For a brief history of the Lower Ninth Ward: Landphair, J. (2007). "'The forgotten people of New Orleans': community, vulnerability, and the Lower Ninth Ward." *The Journal of American History* 94(3), 837-845.

24   Bouie, J. (2015, August 23). "Where Black Lives Matter Began." *Slate*. http://www.slate.com/articles/news_and_politics/politics/2015/08/hurricane_katrina_10th_anniversary_how_the_black_lives_matter_movement_was.html.

The Lady in the Purple Outfit

25   For an in-depth sociological perspective of Katrina read: Bankston III, et al. (2010). *The Sociology of Katrina: Perspectives on a Modern Catastrophe*. Rowman & Littlefield Publishers.

26   Berkes, H. (2005, October 13). "Hurricane debris piles up in Louisiana." NPR. https://www.npr.org/templates/story/story.php?storyId=4957616.

27   Schleifstein, M. (2012, February 27). "New Orleans trees show nation's steepest dropoff." *The Times-Picayune*. https://www.nola.com/news/environment/article_3b928a4d-808c-51e6-9be3-4d5c12cbe0ff.html#:~:text=%E2%80%9CWhen%20you%20see%20it%20in,trees%20during%20and%20after%20Katrina.

28   For more on the Katrina induced oil spills see: Pine, J. C. (2006). "Hurricane Katrina and oil spills: Impact on coastal and ocean environments." *Oceanography* 19(2), 37-39.

29   I use the term "community" in a broad sense throughout this book. Community can be based on geographic proximity, formal jurisdictional boundaries, social relationships, or cultural ties.

30   The Data Center. (2006). "Current housing unit damage estimates." The Data Center. https://gnocdc.s3.amazonaws.com/reports/Katrina_Rita_Wilma_Damage_2_12_06___revised.pdf.

31   Plyer, A. (2016, August 26). "Facts for features: Katrina impact." The Data Center. https://www.datacenterresearch.org/data-resources/katrina/facts-for-impact.

A Heckuva Job!

32   For more on the distinction in terms: Quarantelli, E. L. (2005, September). "Catastrophes are different from disasters: some implications for crisis planning and managing drawn from Katrina." In Online forum and essays Social Science Research Council. https://items.ssrc.org/understanding-katrina/catastrophes-are-different-from-disasters-some-implications-for-crisis-planning-and-managing-drawn-from-katrina.

33   National Hurricane Center, (n.d.). "Hurricanes in History." National Oceanic and Atmospheric Administration. http://www.nhc.noaa.gov/outreach/history/#katrina.

34   Seed, R. B., et al.(2005). *"Preliminary report on the performance of the New Orleans levee systems in Hurricane Katrina on August 29, 2005."* Independent Levee Hurricane Katrina Investigation Team, Preliminary Report. https://www.berkeley.edu/news/media/releases/2005/11/leveereport_prelim.pdf.

       For more on the New Orleans levees and the Corps of Engineers read: Rosenthal, S. (2020). *Words Whispered in Water: Why the Levees Broke in Hurricane Katrina.* Mango Publishing.

35   For a detailed accounting of the response to Katrina and the levee failure see: Brinkley, D. (2006). *The Great Deluge: Hurricane Katrina, New Orleans, and the Mississippi Gulf Coast.* William Morrow. And Horne, J. (2008). *Breach of Faith: Hurricane Katrina and The Near Death of a Great American City.* Random House Incorporated.

36   Sullivan, L. (2005, September 23). "How New Orleans' evacuation plan fell apart." NPR: https://www.npr.org/templates/story/story.php?storyId=4860776.

37   Trainor, J., and Barsky, L. (2011). "Reporting for Duty? A Synthesis of Research on Role Conflict, Strain, and Abandonment Among Emergency Responders during Disasters and Catastrophes." Disaster Research Center.

38   Scharf, P. and Phillippi, S. (2015, August 27). "The New Orleans Police Department was troubled long before Hurricane Katrina." *The Conversation.* https://theconversation.com/the-new-orleans-police-department-was-troubled-long-before-hurricane-katrina-46381.

39   For more on these cases: N.A. (2016, April 20). "Danziger Bridge officers sentenced: 7 to 12 years for shooters, cop in cover-up gets 3." *The Times-Picayune.* https://www.nola.com/news/crime_police/article_5ee6da31-48c1-5740-897e-4fe177c66930.html. And Robertson, C. (2010, December 9). "Jury Convicts 3 Officers in Post-Katrina Death." *New York Times.* https://www.nytimes.com/2010/12/10/us/10katrina.html.

40 Brinkley, D. (2006). *The Great Deluge: Hurricane Katrina, New Orleans, and the Mississippi Gulf Coast.* William Morrow.

41 Smith, M. (2005, November 24). "The Storm." PBS: Frontline. https://www.pbs.org/wgbh/pages/frontline/storm/etc/script.html.

42 Pasternak, J. and Neubauer, C. (2005, September 4). "Political Connection Led Chief to FEMA." Los Angeles Times. https://www.latimes.com/archives/la-xpm-2005-sep-04-na-brown4-story.html.

43 His prior employment included serving as the commissioner of the International Arabian Horse Association and general counsel for various companies in the energy sector.

44 Think Progress. (2005, September 7). "Katrina Timeline." *Think Progress.* https://thinkprogress.org/katrina-timeline-90ec8a71fb99.

45 Jackson Free Press. (2005, September 4). "Time-Picayune's Open Letter to the President." Jackson Free Press. https://www.jacksonfreepress.com/news/2005/sep/04/times-picayunes-open-letter-to-the-president/.

46 Pao, M. (2015, August 27). "Swept up in the storm: Hurricane Katrina's key players, then and now." NPR. https://www.npr.org/2015/08/27/434385285/swept-up-in-the-storm-hurricane-katrinas-key-players-then-and-now.

47 Moynihan, D. P. (2015). "Collaboration amid crisis: The Department of Defense during Hurricane Katrina." Casoteca of Public Management.

48 "How NOLA Superdome made it back after Katrina." [Video file]. (2013, February 2). https://www.youtube.com/watch?v=I-YotwUv4bA&ab_channel=LouisianaPublicBroadcasting.

49 Suarez, R. (2005, September 9). "Public Opinion After Katrina." [Radio]. PBS. https://www.pbs.org/newshour/show/public-opinion-after-katrina.

50 Serrano, R. A. (2005, September 4). "Bush Promises Improved Relief Effort, Adds Troops." *Los Angeles Times.* https://www.latimes.com/archives/la-xpm-2005-sep-04-na-bush4-story.html.

51 N.A. (2015, August 29). "How citizens turned into saviors after Katrina struck." *CBS This Morning.* https://www.cbsnews.com/news/remembering-the-cajun-navy-10-years-after-hurricane-katrina.

52 Horwitz, S. (2009). "Best responders: Post-Katrina innovation and improvisation by Wal-Mart and the US Coast Guard." *Innovations: Technology, Governance, Globalization* 4(2), 93-99.

53 Quarantelli, E. L. (2005, September). "Catastrophes are different from disasters: some implications for crisis planning and managing drawn from Katrina." In Online forum and essays Social Science Research Council. https://items.ssrc.org/understanding-katrina/catastrophes-are-different-from-disasters-some-implications-for-crisis-planning-and-managing-drawn-from-katrina.

54 Mustian, J. (2015, August 29). "A decade later, counting Katrina's dead remains an imperfect science at best." *The Advocate.* https://www.theadvocate.com/baton_rouge/news/article_c3330e57-f853-517e-ac4b-7589ae82b0c0.html.

55   Bialik, C. (2015, August 26). "We still don't know how many people died because of Katrina." *FiveThirtyEight*. https://fivethirtyeight.com/features/we-still-dont-know-how-many-people-died-because-of-katrina.

56   Brunkard, J., Namulanda, G., and Ratard, R. (2008). "Hurricane Katrina Deaths, Louisiana, 2005." *Disaster Medicine and Public Health Preparedness* 2(4), 215-223.

57   For a longer historical context of Katrina read: Horowitz, A. (2020). *Katrina: A History, 1915-2015*. Harvard University Press.

58   Preparing for a catastrophe: The Hurricane Pam exercise. (2006, January 24). Hearing before the Committee on Homeland Security and Governmental Affairs US Senate. 109 Congress. https://biotech.law.lsu.edu/blaw/FEMA/CHRG-109shrg26749.pdf.

59   Townsend, F. F. (2006). "The Federal Response to Hurricane Katrina: Lessons Learned." Washington, DC: The White House.

Do-It-Yourself Recovery

60   Pope, J. (2008, February 23). "Students came to help, return to learn." *The Times-Picayune*. https://www.nola.com/news/article_d59a1e1d-0844-5af3-bc57-198669d4de8b.html.

61   Shapiro, I. and Sherman, A. (2005, September 19). "Essential Facts About The Victims of Hurricane Katrina." Center on Budget and Policy Priorities. https://www.cbpp.org/research/essential-facts-about-the-victims-of-hurricane-katrina.

62   AP. (2008, August 26). "Hurricane recovery confronts low literacy rate." NBC News. https://www.nbcnews.com/id/wbna26413788.

63   Jenkins, P., Lambeth, T., Mosby, K., and Van Brown, B. (2015). "Local Nonprofit Organizations in a Post-Katrina Landscape: Help In a Context of Recovery." *American Behavioral Scientist* 59(10), 1263-1277.

64   Montano, S. L. (2014). "Formation and Lifespans of Emergent Recovery Groups in Post-Katrina New Orleans." Thesis. North Dakota State University.

65   For an in-depth look at the recovery in New Orleans read: Rivlin, G. (2016). *Katrina: After the Flood*. Simon & Schuster.

66   Gould, L. A. (2014). "A Conceptual Model of The Individual and Household Recovery Process: Examining Hurricane Sandy." Thesis. North Dakota State University.

67   The Federal Reserve. (2016). "Report on the Economic Well-Being of US Households in 2015." The Federal Reserve. https://www.federalreserve.gov/2015-report-economic-well-being-us-households-201605.pdf.

68   For more on the National Flood Insurance Program see: Montano, S. and Savitt, A. (2018, August 1). "The Cost of Flood Insurance Is a Price Worth Paying." *City Lab*. https://www.bloomberg.com/news/articles/2018-08-01/the-cost-of-flood-insurance-is-one-america-must-shoulder.

69   Congressional Research Service. (2020, October 2). National Flood In-

surance Program Borrowing Authority. https://fas.org/sgp/crs/homesec/IN10784.pdf.

70 Insurance Information Institute. (n.d.). National Flood Insurance Program. https://www.iii.org/fact-statistic/facts-statistics-flood-insurance#:~:text=The%202018%20Insurance%20Information%20Institute,had%20the%20coverage%20in%202016.

71 Cornish, A. (2019, July 9). "Why Only 13% Of California Homeowners Have Earthquake Insurance." [Radio]. NPR. https://www.npr.org/2019/07/09/739999709/why-only-13-of-california-homeowners-have-earthquake-insurance.

72 Schalch, K. (2005, September 9). "Many Katrina Flooded Homes Had No Insurance." [Radio]. NPR. https://www.npr.org/templates/story/story.php?storyId=4838689.

73 Richard Campanella. (2010). *Bienville's Dilemma: A Historical Geography of New Orleans.* Garrett County Press.

74 For more on the Stafford Act and disaster policy read: Sylves, R. T. (2019). *Disaster Policy and Politics: Emergency Management and Homeland Security.* CQ Press.

75 Sylves, R. T. (2019). *Disaster Policy and Politics: Emergency Management and Homeland Security.* CQ Press.

76 Presser, L. (2019, July 15). "Their Family Bought Land One Generation After Slavery." *ProPublica.* https://features.propublica.org/black-land-loss/heirs-property-rights-why-black-families-lose-land-south/?utm_source=twitter&utm_medium=social.

77 Ydstie, J. (2008, April 28). "No Title? No Easy Access to Post-Katrina Aid." NPR. https://www.npr.org/templates/story/story.php?storyId=90005954.

78 Calder, C. (2016, August 17). "How much money can you expect from FEMA? Disaster grants sure to disappoint, analysis finds." *The Advocate.* http://www.theadvocate.com/louisiana_flood_2016/article_22c86fe0-64cd-11e6-9bb2-07f95d36ee28.html.

79 Food Research & Action Center. (2018). "The FRAC Advocate's Guide to the Disaster Supplemental Nutrition Assistance Program (D-SNAP)." https://frac.org/wp-content/uploads/d-snap-advocates-guide-1.pdf.

80 Shapiro, I. (2005, September 27). "Benefit Levels for Unemployed Hurricane Victims are Too Low." Center on Budget and Policy Priorities. https://www.cbpp.org/research/benefit-levels-for-unemployed-hurricane-victims-are-too-low.

81 Goldstein, Z. (2019, August 27). "New Data: Why 800,000 Applicants Were Denied Federal Disaster Assistance Loans." The Center for Public Integrity. https://publicintegrity.org/environment/new-data-reveals-why-800000-applicants-were-denied-federal-disaster-assistance-loans.

82 Weiss, M. (2010, August 24). "AP Investigation: Katrina a tale of SBA failure." AP. http://archive.boston.com/business/articles/2010/08/24/ap_investigation_katrina_a_tale_of_sba_failure/?page=1.

83 Wu, C. (2020, August 7). "Reparations, Race, and Reputation in Credit:

Rethinking the Relationship Between Credit Scores and Reports with Black Communities." *Medium.* https://medium.com/@cwu_84767/reparations-race-and-reputation-in-credit-rethinking-the-relationship-between-credit-scores-and-852f70149877.

84 Frank, T. (2020, July 2). "Disaster Loans Entrench Disparities in Black Communities." *Scientific American.* https://www.scientificamerican.com/article/disaster-loans-entrench-disparities-in-black-communities.

85 For more on individual recovery and grassroots efforts post-Katrina read: Gratz, R. B. (2015). *We're Still Here Ya Bastards: How the People of New Orleans Rebuilt Their City.* Bold Type Books.

86 For more on race and environmental justice in the context of Katrina read: Bullard, R., and Wright, B. (2009). *Race, Place, and Environmental Justice after Hurricane Katrina.* Boulder, CO: Westview.

87 N.A. (2009, January 6). "The Road Home Program." [Video]. PBS Frontline. https://www.pbs.org/wgbh/pages/frontline/katrina/fail/roadhome.html.

88 Love, N. P. (2008). "The Louisiana Road Home Program: Federal Aid for State Disaster Housing Assistance Programs." Congressional Research Service Report for Congress. https://www.everycrsreport.com/reports/RL34410.html.

89 Hammer, D. (2015, August 23). "Examining post-Katrina Road Home program." *The Advocate.* http://www.theadvocate.com/baton_rouge/news/article_f9763ca5-42ba-5a62-9935-c5f7ca94a7c4.html.

90 Fletcher, M. A. (2011, July 6). "HUD to pay $62 million to La. Homeowners to settle Road Home lawsuit." *Washington Post.* https://www.washingtonpost.com/business/economy/hud-to-pay-62-million-to-la-homeowners-to-settle-road-home-lawsuit/2011/07/06/gIQAtsFN1H_story.html.

91 McClendon, R. (2014, March 14). "Road Home rules have changed, many said to owe money may not after all." *The Times-Picayune.* https://www.nola.com/news/politics/article_8a8d298b-8c8e-573c-88ea-f8bf9b2310d1.html.

92 Brunker, M. (2008, February 14). CDC test confirm FEMA trailers are toxic. NBC News. https://www.nbcnews.com/id/wbna23168160.

Green Dots

93 Flaherty, J. (2008, September 11). "Three Years After Katrina, A City Still Threatened." *The Indypendent.* https://indypendent.org/2008/09/three-years-after-katrina-a-city-still-threatened.

94 For a longer discussion of the term "build back better" see: Chmutina, K. and Cheek, W. (2021, January 22). "Build Back Better for Whom? How Neoliberalism (Re)creates Disaster Risks." Current Affairs. https://www.currentaffairs.org/2021/01/build-back-better-for-whom-how-neoliberalism-recreates-disaster-risks.

95 Klein, N. (2007). *The Shock Doctrine: The Rise of Disaster Capitalism.* Macmillan.

96 Klein, N. (2017, July 6). "How power profits from disaster." *The Guardian.*

https://www.theguardian.com/us-news/2017/jul/06/naomi-klein-how-power-profits-from-disaster.

97   For more on market-driven governance in Post-Katrina New Orleans see: Adams, V. (2013). *Markets of Sorrow, Labors of Faith: New Orleans in the Wake of Katrina*. Duke University Press.

98   Mock, B. (2015, August 27). "Why Louisiana Fought Low-Income Housing in New Orleans after Katrina." *City Lab*. https://www.citylab.com/equity/2015/08/why-louisiana-fought-low-income-housing-in-new-orleans-after-katrina/401939.

99   Babington, C. (2005, September 10). "Some GOP Legislators Hit Jarring Notes in Addressing Katrina." The Washington Post. https://www.washingtonpost.com/archive/politics/2005/09/10/some-gop-legislators-hit-jarring-notes-in-addressing-katrina/685fa514-1b19-4893-934b-9200d1f4d608/.

100 Klein, N. (2018, September 21). "There's nothing natural about Puerto Rico's disaster." *The Intercept*. https://theintercept.com/2018/09/21/puerto-rico-hurricane-maria-disaster-capitalism.

101 Lamb, Z. (2020). "Connecting the Dots: The Origins, Evolutions, and Implications of the Map that Changed Post-Katrina Recovery Planning in New Orleans." *Louisiana's Response to Extreme Weather*, (65-91). Springer, Cham.

102 Hennick, C. (n.d.). "A Tale of Two Neighborhoods." USGBC. https://plus.usgbc.org/a-tale-of-two-neighborhoods.

103 Simerman, J. (2020, August 29). "New Orleans's Lower 9th Ward is Still Reeling from Hurricane Katrina's Damage 15 Years Later." *The Advocate*. https://www.nola.com/news/katrina/article_a192c350-ea0e-11ea-a863-2bc584f57987.html.

104 Robertson, C. (2014, February 12). "Nagin Guilty of 20 Counts of Bribery." New York Times. https://www.nytimes.com/2014/02/13/us/nagin-corruption-verdict.html.

105 Russell, G. and Varney, J. (2005, December 29). "From blue tarps to debris removal, layers of contractors drive up the cost of recovery, critics say." *The Times-Picayune*. https://www.pulitzer.org/winners/times-picayune.

Bootstraps

106 Kuligowski, E. D., et al. (2014). Final report, National Institute of Standards and Technology (NIST) Technical Investigation of the May 22, 2011, Tornado in Joplin, Missouri. https://www.nist.gov/el/disaster-resilience/joplin-missouri-tornado-2011.

107 Mensah, G. A., et al. (2005). "When Chronic Conditions Become Acute: Prevention and Control of Chronic Diseases and Adverse Health Outcomes During Natural Disasters." *Preventing Chronic Disease* 2 (Special No.A04).

108 Krug, E. G., et al. (1998). "Suicide after natural disasters." *New England Journal of Medicine* 338(6), 373-378.

109 Sety, M., James, K., and Breckenridge, J. (2014). "Understanding the Risk of

Domestic Violence During and Post Natural Disasters: Literature Review." *Issues of Gender and Sexual Orientation in Humanitarian Emergencies*, (99-111). Springer, Cham.

110   Montano, S., and Savitt, A. (2016). "Rethinking Our Approach to Gender and Disasters: Needs, Responsibilities, and Solutions." *Journal of Emergency Management* (Weston, Mass.) 14(3), 189-199.

111   Rudowitz, R., Rowland, D., and Shartzer, A. (2006). "Health Care In New Orleans Before And After Hurricane Katrina: The storm of 2005 exposed problems that had existed for years and made solutions more complex and difficult to obtain." *Health Affairs* 25(Suppl. 1).

112   CBS/AP. (2008, February 19). "New Orleans Facing Mental Health Crisis." CBS News. https://www.cbsnews.com/news/new-orleans-facing-mental-health-crisis.

113   Brasted, C. (2018, April 5). "New Orleans Suicides Skyrocketed After Katrina. Here's Where We Are Now." *The Times-Picayune*. https://www.nola.com/health/index.ssf/2018/03/new_orleans_suicide_data.html.

114   For a history of emergency management in the United States see: Rubin, C. B. (Ed.). (2019). *Emergency Management: The American Experience*. Routledge.

115   Comerio, M. C. (1998). *Disaster Hits Home: New Policy for Urban Housing Recovery*. University of California Press.

116   For more on long-term community recovery see: Alesch, D. J., Arendt, L. A., and Holly, J. N. (2009). *Managing For Long-Term Community Recovery in The Aftermath of Disaster*. Public Entity Risk Institute.

117   For more on how multi-stakeholder recovery is approached in the US see: Smith, G. (2012). *Planning for Post-Disaster Recovery: A Review of The United States Disaster Assistance Framework*. Island Press.

Commemorative Snow Globes

118   Corporation for National & Community Service (2007). "Number of Volunteers in Year 2 of Katrina Recovery Exceeds Historic 1st Year." https://www.nationalservice.gov/pdf/07_0820_katrina_volunteers_respond.pdf. (Last accessed on July 14, 2020).

119   Montano, S. (2019). "Disaster volunteerism as a contributor to resilience." *The Routledge Handbook of Urban Resilience*. Routledge.

120   Plyer, A. (2016, August 26). "Facts For Features: Katrina Impact." Nola Data Center. https://www.datacenterresearch.org/data-resources/katrina/facts-for-impact.

121   White, J. (2015, May 30). "New Orleans mayor Mitch Landrieu: City 'no longer recovering, no longer rebuilding' 10 years after Katrina." *The Times-Picayune*. https://www.nola.com/news/politics/article_08510590-85bf-53e5-88c9-cc8d380e10cf.html.

122   Ehrenfeucht, R., and Nelson, M. (2013). "Young Professionals as Ambivalent Change Agents in New Orleans After The 2005 Hurricanes." *Urban Studies* 50(4), 825-841.

PART TWO

Mitigation: Preventing Disaster

123 Koenig, S. (2018, March 2). "Nor'easter clips Maine, knocking out power, flooding roads and canceling flights." *Bangor Daily News*. http://bangor-dailynews.com/2018/03/02/weather/mainers-brace-for-flooding-power-outages-along-the-coast.

124 Breton, R. (2018, March 3). "Streets washed out in latest round of flooding." *News Center Maine*. https://www.newscentermaine.com/article/news/local/streets-washed-out-in-latest-round-of-flooding/97-525161845.

125 Wright, V. M. (2018, August). "Sweet, Fleeting Season." *DownEast Magazine*. https://downeast.com/our-towns/sweet-fleeting-season.

126 FEMA. (2018). Maine Severe Storm and Flooding (DR-4367-ME). https://www.fema.gov/disaster/4367.

Creating Disaster

127 Unless otherwise noted this history of Saco comes from N.A. (1976) "A Folklore History of Saco." House of Falmouth.

128 Baker, E. (1987). "Archaeology in the Saco area." https://www.sacomaine.org/residents/city_history/archaeology.php.

129 Hardiman, T. (2018). "A history of the factory island mill district." https://www.sacomaine.org/residents/city_history/factory_island.php.

130 Graham, G. (2018, April 15). "After losing ground for decades, Camp Ellis feeling helpless." *Portland Press Herald*. https://www.centralmaine.com/2018/04/15/after-losing-ground-for-decades-camp-ellis-feeling-helpless.

131 USACE. (2013). "Section 111 Shore Damage Mitigation Project Draft Decision Document and Environmental Assessment: Saco River and Camp Ellis Beach Saco, Maine." https://www.nae.usace.army.mil/Portals/74/docs/Topics/CampEllis/MainReportDRAFT.pdf.

132 For a more complete history of the jetty in Camp Ellis see: Cervone, E. (2003). "An Engineering, Economic, and Political Approach to Beach Erosion Mitigation and Harbor Development: A Review of the Beach Communities of Camp Ellis, Maine, Wells, Maine, and Cape May, New Jersey." Electronic Theses and Dissertations. The University of Maine.

133 Wright, V. M. (2018, August). "Sweet, Fleeting Season." *DownEast Magazine*. https://downeast.com/our-towns/sweet-fleeting-season.

134 For extensive discussion on risk see: Wisner, B., et al. (2004). *At Risk: Natural Hazards, People's Vulnerability and Disasters*. Psychology Press. And Tierney, K. (2014). *The Social Roots of Risk: Producing Disasters, Promoting Resilience*. Stanford University Press.

135 Scully, J. (2000). *The Night Old Orchard Died: The Great Old Orchard Fire of 1907*. Doodlebug Publishing.

136 Schreiber, L. (2017, September 1). "Deal for Saco Island land sets stage for

$40M project." *Mainebiz*. https://www.mainebiz.biz/article/deal-for-saco-island-land-sets-stage-for-40m-project.

137　National Ocean Service. (2020, July 16). "What percentage of the American population lives near the coast?" National Oceanic and Atmospheric Administration. https://oceanservice.noaa.gov/facts/population.html.

138　For an extensive investigation of urban risk see: Dawson, A. (2017). *Extreme cities: The Peril and Promise of Urban Life in the Age of Climate Change*. Verso Books.

139　Comerio, M. C. (1998*). Disaster Hits Home: New Policy for Urban Housing Recovery*. University of California Press.

140　Townsend, F. F. (2006). "The Federal Response to Hurricane Katrina: Lessons Learned." Washington, DC: The White House. https://tools.niehs.nih.gov/wetp/public/hasl_get_blob.cfm?ID=4628.

141　Dynes, R. R. (1999). "The dialogue between Voltaire and Rousseau on the Lisbon earthquake: The emergence of a social science view." Preliminary Paper #293. Disaster Research Center. University of Delaware.

142　Kälin, W. (2011, June 6). "A Human Rights-Based Approach to Building Resilience to Natural Disasters." Brookings. https://www.brookings.edu/research/a-human-rights-based-approach-to-building-resilience-to-natural-disasters.

143　White, G. F. (1945). "Human Adjustment to Floods." University of Chicago Department of Geography Research Paper No. 29. Chicago: University of Chicago Department of Geography.

144　O'Keefe, P., Westgate, K., and Wisner, B. (1976). "Taking the naturalness out of natural disasters." *Nature*. 260, 566–567. https://doi.org/10.1038/260566a0.

The Fight for Camp Ellis

145　USACE. (2013). "Section 111 Shore Damage Mitigation Project Draft Decision Document and Environmental Assessment: Saco River and Camp Ellis Beach Saco, Maine." https://www.nae.usace.army.mil/Portals/74/docs/Topics/CampEllis/MainReportDRAFT.pdf.

146　Wright, V. M. (2018, August). "Sweet, Fleeting Season." *DownEast Magazine*. https://downeast.com/our-towns/sweet-fleeting-season.

147　Wright, V. M. (2018, August). "Sweet, Fleeting Season." *DownEast Magazine*. https://downeast.com/our-towns/sweet-fleeting-season.

148　Cervone, E. (2003). "An Engineering, Economic, and Political Approach to Beach Erosion Mitigation and Harbor Development: A Review of the Beach Communities of Camp Ellis, Maine, Wells, Maine, and Cape May, New Jersey." Electronic Theses and Dissertations. The University of Maine.

149　Birkland, T. A. (1998). "Focusing Events, Mobilization, and Agenda Setting." *Journal of Public Policy* 18(1), 53-74.

150　Wright, V. M. (2018, August). "Sweet, Fleeting Season." *DownEast Magazine*. https://downeast.com/our-towns/sweet-fleeting-season.

151 Wright, V. M. (2018, August). "Sweet, Fleeting Season." *DownEast Magazine*. https://downeast.com/our-towns/sweet-fleeting-season.

Disaster Denialism

152 Carlowicz, M. (2018, September 12). "Watery heatwave cooks the Gulf of Maine." NASA's Earth Observatory. https://climate.nasa.gov/news/2798/watery-heatwave-cooks-the-gulf-of-maine.

153 Albeck-Ripka, L. (2018, June 21). "Climate Change Brought a Lobster Boom. Now It Could Cause a Bust." *New York Times*. https://www.nytimes.com/2018/06/21/climate/maine-lobsters.html.

154 Kahn, B. (2015, October 29). "Climate change is decimating cod in the Gulf of Maine." Climate Central. https://www.climatecentral.org/news/climate-change-decimating-cod-maine-19617.

155 Garcia-Navarro, L. (2019, October 6). "The Gulf of Maine Is Warming, And Its Whales Are Disappearing". NPR. https://www.npr.org/2019/10/06/766401296/the-gulf-of-maine-is-warming-and-its-whales-are-disappearing.

156 Gulf of Maine Research Institute (n.d.). "Sea Level Rise in Maine." https://gmri.maps.arcgis.com/apps/Cascade/index.html?appid=a043f621616d4f8b8088f0117796896d.

157 Berardelli, J. (2019, July 8). "How climate change is making hurricanes more dangerous." Yale Climate Connections. https://yaleclimateconnections.org/2019/07/how-climate-change-is-making-hurricanes-more-dangerous.

158 For more see: Leonard, C. (2020). *Kochland: The Secret History of Koch Industries and Corporate Power in America*. Simon & Schuster.

159 N.A. (2017, September 11). "Bossert says White House takes climate change 'we observe' seriously, not its cause." [Video]. *Washington Post*. https://www.washingtonpost.com/video/politics/bossert-says-white-house-takes-climate-change-we-observe-seriously-not-its-cause/2017/09/11/d935b696-9723-11e7-af6a-6555caaeb8dc_video.html.

160 Montano, S. (2018, November 16). "Trump's Response To The California Wildfires Isn't Just Inaccurate. It's Dangerous." *Huffington Post*. https://www.huffpost.com/entry/opinion-california-wildfires-cause-disaster_n_5bec49cbe4b0783e0a1ef488.

161 Pierre-Louis, K. (2018, November 12). "Trump's Misleading Claims About California's Fire 'Mismanagement.'" *New York Times*. https://www.nytimes.com/2018/11/12/us/politics/fact-check-trump-california-fire-tweet.html?action=click&module=MoreInSection&pgtype=Article&region=Footer&contentCollection=Climate%20and%20Environment.

162 Pierre-Louis, K. (2018, November 9). "Why Does California Have So Many Wildfires?" *New York Times*. https://www.nytimes.com/2018/11/09/climate/why-california-fires.html.

163 N.A. (2005, September 1). "Bush Insists Help Is On The Way." BBC News. http://news.bbc.co.uk/2/hi/4204754.stm.

164   For more on the distinction between emergencies, disasters, and catastrophes see: Tierney, K. (2019). *Disasters: A Sociological Approach*. John Wiley & Sons.

A Reactive Approach

165   Lindell, M. K., Perry, R. W., Prater, C., and Nicholson, W. C. (2006). "Chapter Seven: Hazard Mitigation" in *Fundamentals of Emergency Management*. Washington, DC, USA: FEMA.

166   For more on the relationship between hazard mitigation and climate adaptation see: Klima, K., and Jerolleman, A. (2017). "A rose by any other name—communicating between hazard mitigation, climate adaptation, disaster risk reduction, and sustainability professionals." *Journal of Environmental Studies and Sciences* 7(1), 25-29.

167   Multi-Hazard Mitigation Council (2019). "Natural Hazard Mitigation Saves: 2019 Report." https://www.nibs.org/projects/natural-hazard-mitigation-saves-2019-report.

168   For a review of this research read: Savitt, A. M. (2020). "The Role of the County Emergency Manager in Disaster Mitigation." Doctoral dissertation. North Dakota State University.

The Future of Camp Ellis

169   For a history of buy-out discussions in Camp Ellis see: Kelley, J.T. and Brothers, L.L. (2007). "Camp Ellis, Maine: A small beach community with a big problem...its jetty." In Kelley, J.T., Pilkey, O.H., & Cooper, J.A.G. (Eds.), America's Most Vulnerable Shorelines: Geological Society of America Special Paper, 460, pp. 1-20.

170   Kennedy, B. and Johnson, C. (2020, February 28). "More Americans see climate change as a priority, but Democrats are much more concerned than Republicans." Pew Research. https://www.pewresearch.org/fact-tank/2020/02/28/more-americans-see-climate-change-as-a-priority-but-democrats-are-much-more-concerned-than-republicans.

171   Benincasa, R. (2019, March 5). "Search The Thousands Of Disaster Buyouts FEMA Didn't Want You To See." NPR. https://www.npr.org/2019/03/05/696995788/search-the-thousands-of-disaster-buyouts-fema-didnt-want-you-to-see.

172   Weber, A. and Moore, R. (2019, September 12). "Going Under: Long Wait Times for Post-Flood Buyouts Leave Homeowners Underwater." National Resource Defense Council. https://www.nrdc.org/resources/going-under-long-wait-times-post-flood-buyouts-leave-homeowners-underwater.

173   Benincasa, R. (2019, March 5). "Search The Thousands Of Disaster Buyouts FEMA Didn't Want You To See." NPR. https://www.npr.org/2019/03/05/696995788/search-the-thousands-of-disaster-buyouts-fema-didnt-want-you-to-see.

174   Anderson, S., Plantinga, A. J., and Wibbenmeyer, M. (2020, December 16). "Inequality in Agency Responsiveness: Evidence from Salient Wildfire Events." *Resources for the Future*. https://www.rff.org/publications/working-papers/inequality-agency-responsiveness-evidence-salient-wildfire-events.

175   Weber, A. (2019, September 17). "Blueprint of a Buyout: Harris County, TX." National Resource Defense Council. https://www.nrdc.org/experts/anna-weber/buyout-case-study-harris-county-texas.

PART THREE

Response: The Disaster

176   Trenberth, K. E. (2018). "Hurricane Harvey Links to Ocean Heat Content and Climate Change Adaptation." *Earth's Future*, 730-744.

177   Heglar, M. A. (2020, October 4). "2020: The Year of the Converging Crises." *Rolling Stone*. https://www.rollingstone.com/politics/political-commentary/2020-crises-wildfires-pandemic-election-climate-crisis-1069907.

178   For more on Hurricane Harvey read: Blake, E. S. and Zelinsky, D. A. (2018, January 23). "National Hurricane Center Tropical Cyclone Report Hurricane Harvey." National Oceanic Atmosphere Administration. https://www.nhc.noaa.gov/data/tcr/AL092017_Harvey.pdf.

179   Chappell, B. (2017, August 28). "National Weather Service Adds New Colors So It Can Map Harvey's Rains." NPR. https://www.npr.org/sections/thetwo-way/2017/08/28/546776542/national-weather-service-adds-new-colors-so-it-can-map-harveys-rains.

180   Kennedy, M. (2018, January 25). "Harvey The 'Most Significant Tropical Cyclone Rainfall Event In U.S. History.'" NPR. https://www.npr.org/sections/thetwo-way/2018/01/25/580689546/harvey-the-most-significant-tropical-cyclone-rainfall-event-in-u-s-history.

181   Jonkman, S. N., Godfroy, M., Sebastian, A., and Kolen, B. (2018). "Brief communication: Loss of life due to Hurricane Harvey." *Natural Hazards and Earth Systems Sciences* 18(4), 1073-1078.

182   N.A. (2018, August). "Hurricane Harvey Report: ShelterBox USA's response in Greater Houston." ShelterBox USA. https://www.shelterboxusa.org/wp-content/uploads/2018/08/ShelterBox-USA-Harvey-Outcomes-Report-2018.pdf.

183   Carlsen, A. and Lai, K. K. R. (2017, September 1). "Where Harvey Hit Hardest Up and Down the Texas Coast." *New York Times*. https://www.nytimes.com/interactive/2017/09/01/us/hurricane-harvey-damage-texas-cities-towns.html.

184   Davies, A. (2017, September 3). "Harvey wrecks up to a million cars in car-dependent Houston." *Wired*. https://www.wired.com/story/harvey-houston-cars-ruined.

185   Mooney, C. (2018, January 8). "Hurricane Harvey was year's costliest U.S. disaster at $125 billion in damages." The Texas Tribune. https://www.texastribune.org/2018/01/08/hurricane-harvey-was-years-costliest-us-disaster-125-billion-damages.

186   For a brief overview of these floods see: N.A. (2018, December 7). "Rained out: Looking back at holidays that ended with flooding in Houston." ABC 13. https://abc13.com/houston-flooding-tax-day-floods-in-memorial-labor/4136696.

A Quick Response

187 For more on the Halifax Explosion read: Kitz, J. (2010). *Shattered City: The Halifax Explosion & the Road to Recovery.* Nimbus Publishing. And Remes, J. A. (2015). *Disaster Citizenship: Survivors, Solidarity, and Power in The Progressive Era.* University of Illinois Press.

188 For more on Samuel Prince and his contribution to disaster research read: Scanlon, T. J. (1988). "Disaster's Little Known Pioneer: Canada's Samuel Henry Prince." *International Journal of Mass Emergencies and Disasters* 6(3), 213–232.

189 Cornell, J. (1982). *The Great International Disaster Book*, 3rd. Ed. Charles Scribner, New York.

190 For more on the history of disaster research see: Tierney, K. (2019). *Disasters: A Sociological Approach.* John Wiley & Sons.

191 For a popular account of one of these initial studies see: Mooallem, J. (2020). This is Chance!: The Shaking of an All-American City, A Voice That Held It Together. Random House. And for an academic perspective: Quarantelli, E. L. (1987). "Disaster studies: An analysis of the social historical factors affecting the development of research in the area." University of Delaware: Disaster Research Center.

192 For a popular overview of prosocial behavior in communities during crisis see: Solnit, R. (2010). *A Paradise Built in Hell: The Extraordinary Communities That Arise in Disaster.* Penguin. And for an academic perspective: Barton, A. H. (1969). *Communities in Disaster: A Sociological Analysis of Collective Stress Situations.* Doubleday & Company, Inc.

193 Tierney, K. (2003). "Disaster beliefs and institutional interests: Recycling disaster myths in the aftermath of 9-11." *Research in Social Problems and Public Policy* 11, 33–51.

194 For more on emergence and convergence: Dynes, R. R. and Quarantelli, E. L. (1980). "Helping behavior in large-scale disasters." In Smith, D. H., and Macaulay, J. (Eds.), *Participation in Social and Political Activities*, (339–354). San Francisco: Jossey-Bass Publishers. And O'Brien, P. and Mileti, D. S. 1992. "Citizen Participation in Emergency Response Following the Loma Prieta Earthquake." *International Journal of Mass Emergencies and Disasters* 10(1): 71–89.

195 Barton, A. H. (1969). *Communities in Disaster: A Sociological Analysis of Collective Stress Situations.* Doubleday & Company, Inc.

196 Beyerlein, K., and Sikkink, D. (2008). "Sorrow and Solidarity: Why Americans Volunteered for 9/11 Relief Efforts." *Social Problems* 55(2), 190–215.

197 Twigg, J., and Mosel, I. (2017). Emergent groups and spontaneous volunteers in urban disaster response. *Environment and Urbanization*, 29(2), 443–458.

198 Dennis S. Mileti. 1992. "Citizen Participation in Emergency Response Following the Loma Prieta Earthquake." *International Journal of Mass Emergencies and Disasters* 10(1): 71–89.

199 Dynes, R. R., Quarantelli, E. L., and Wenger, D. (1988). "The Organiza-

tional and Public Response to The September 1985 Earthquake in Mexico City, Mexico." Final Report #35. Newark, DE: Disaster Research Center.

200 Twigg, J., and Mosel, I. (2017). "Emergent Groups and Spontaneous Volunteers in Urban Disaster Response." *Environment and Urbanization* 29(2), 443-458.

201 For more on the research in Texas that is described in this chapter see: Montano, S. (2017). "A foundation for factors that explain volunteer engagement in response and recovery: The case of flooding in East Texas 2016." Doctoral Dissertation. North Dakota State University, North Dakota, USA.

202 For a review of the disaster volunteerism literature see: Montano, S. (2019). Disaster volunteerism as a contributor to resilience. *The Routledge Handbook of Urban Resilience*. Routledge: p. 217.

203 For more on how we define response read: Tierney, K. J., Lindell, M. K., and Perry, R. W. (2001). *Facing The Unexpected: Disaster Preparedness and Response in The United States*. Joseph Henry Press.

204 For more on the basics of the phase of response read: Phillips, B. D., Neal, D. M., and Webb, G. (2011). *Introduction to Emergency Management*. CRC Press.

205 For an overview of disaster volunteerism read: Phillips, B. D. (2020). *Disaster Volunteers: Recruiting and Managing People Who Want to Help*. Butterworth-Heinemann.

A Warning

206 This history of the Galveston storm comes from Larson, E. (2011). *Isaac's Storm: A Man, A Time, and The Deadliest Hurricane in History*. Vintage.

207 As with many disasters the Galveston Hurricane death toll is disputed. Estimates generally range from six thousand to twelve thousand. Regardless of where in this range the true death toll falls it would still be the deadliest US hurricane as of this writing. For more on an effort to create a 1900 Storm Victim Database see: https://www.galvestonhistorycenter.org/digitized-collections/1900-storm-victim-database.

208 There is an expansive literature on disaster warnings. For a basic overview see: Drabek, T. E. (1999). "Understanding disaster warning responses." *The Social Science Journal* 36(3), 515-523.

209 For more on the possibilities and challenges of earthquake early warnings see: Minson, S. E., et al. (2019). "The Limits of Earthquake Early Warning Accuracy and Best Alerting Strategy." *Scientific Reports* 9(1), 1-13.

210 Ritchie, H. (2014). "Natural Disasters." Published online at OurWorldInData.org. https://ourworldindata.org/natural-disasters#number-of-deaths-from-natural-disasters.

211 For a brief summary of changes to New Orleans's evacuation plans see: Reckdahl, K. (2014, July 9). "Why New Orleans' Katrina Evacuation Debacle Will Never Happen Again." Next City. https://nextcity.org/daily/entry/new-orleans-evacuation-hurricane-katrina-will-never-happen-again.

212 Haynes, K., et al. (2009). "'Shelter-in-place' vs. evacuation in flash floods." *Environmental Hazards* 8(4), 291-303.

213 For more on the Gustav evacuation see: N.A. (2008, September 1). "Gustav Weakens to Category 1 as Wind, Rain Batters Gulf Coast." PBS News Hour. https://www.pbs.org/newshour/world/weather-july-dec08-gustav_09-01.

214 For more on Gustav's impacts: Rogers, S. (2008, September 1). "Hurricane Gustav hits land southwest of New Orleans." *Al.com.* http://blog.al.com/weather/2008/09/hurricane_gustav_hits_land_sou.html.

215 For a more in-depth analysis see: Cutter, S. L., and Smith, M. M. (2009). "Fleeing from the Hurricane's Wrath: Evacuation and the two Americas." *Environment: Science and Policy for Sustainable Development* 51(2), 26-36.

216 For more on the Evacuteer program and the opportunities for volunteer involvement in mass evacuations see: Hess, D. B., Conley, B. W., and Farrell, C. M. (2014). "Enhancing Capacity for Emergency Evacuation through Resource Matching and Coordinated Volunteerism." *International Journal of Transportation Science and Technology* 2(3), 33-52.

217 Savitt, A. (2015). "An evaluation of the protective action decision model using data from a train derailment in Casselton, North Dakota." Thesis. North Dakota State University.

218 Lindell, M. K., and Perry, R. W. (2012). "The protective action decision model: theoretical modifications and additional evidence." *Risk Analysis: An International Journal* 32(4), 616-632.

219 For more on how these factors may influence evacuation decisions see: Lazo, J. K., et al. (2015). "Factors Affecting Hurricane Evacuation Intentions." *Risk Analysis* 35(10), 1837-1857.

220 For an example of how financial barriers affected evacuation decisions during Hurricane Katrina see: Elder, K., et al. (2007). "African Americans' Decisions Not to Evacuate New Orleans Before Hurricane Katrina: A Qualitative Study." *American Journal of Public Health* 97 (Suppl. 1), S124-S129.

221 An important example are nursing home residents: Dosa, D., et al. (2012). "To Evacuate or Shelter in Place: Implications of Universal Hurricane Evacuation Policies on Nursing Home Residents." *Journal of the American Medical Directors Association* 13(2), 190-e1.

222 For a basic review of shelter behavior see: Quarantelli, E. L. (1995). "Patterns of sheltering and housing in US disasters." *Disaster Prevention and Management.*

223 http://www.weathermate.net/2016/04/20/blog/20-mind-boggling-photographs-of-the-catastrophic-floods-in-texas.

A Planned Response

224 For more on the recent history of flooding in Southeast Texas see: Freedman, A. and Samenow, J. (2019, September 20). "Flooded again: Climate change is making flooding more frequent in Southeast Texas." *Washington Post.* https://www.washingtonpost.com/weather/2019/09/20/flooded-again-

climate-change-is-making-flooding-more-frequent-southeast-texasthanks-part-climate-change.

225 For more on the various factors that have contributed to increasing flood risk in Texas see: *The Texas Tribune* and *ProPublica*. (2016, December 6). "Boomtown, Flood Town." *The Texas Tribune*. https://www.texastribune.org/2016/12/06/houston-flooding-boomtown-flood-town-plain-text.

226 For a review of how hurricanes are affected by climate change see: Romm, J. (2018). *Climate Change: What Everyone Needs to Know*. New York: Oxford University Press.

227 *The Texas Tribune* and *ProPublica*. (2016, December 6). "Boomtown, Flood Town." *The Texas Tribune*. https://www.texastribune.org/2016/12/06/houston-flooding-boomtown-flood-town-plain-text.

228 Bogost, I. (2017, August 28). "Houston's Flood Is a Design Problem." *The Atlantic*. https://www.theatlantic.com/technology/archive/2017/08/why-cities-flood/538251.

229 For more on the storm itself: Lanza, M. (2016, April 26). "Houston's Tax Day flooding put into historical perspective." Space City Weather. https://spacecityweather.com/houstons-flooding-review.

230 For more on flooding in Greenspoint see: Watkins, M. and Formby, B. (2017, September 7). "In battered Houston apartments, residents wonder whether to stay." *The Texas Tribune*. https://www.texastribune.org/2017/09/07/battered-houston-apartment-complex-residents-wonder-whether-they-can-s/ And Hauslohner, A. (2017, September 5). "Recovering from Harvey when 'you already live a disaster every day of your life.'" *Washington Post*. https://www.washingtonpost.com/national/recovering-from-harvey-when-you-already-live-a-disaster-every-day-of-your-life/2017/09/05/40a07e10-9247-11e7-8754-d478688d23b4_story.html.

231 For numbers on damage see: Smith, A., et al. (2016). "U.S. billion-dollar weather & climate disasters 1980-2016." https://www.ncdc.noaa.gov/billions/events.pdf. And Federal Emergency Management Agency. (2016a). "Texas—Severe Storms and Flooding: FEMA-4269-DR." https://www.fema.gov/disaster/4269.

232 For the history of emergency management in the United States read: Rubin, C. B. (Ed.). (2019). *Emergency Management: The American Experience*. Routledge.

233 For more on the emergence of emergency management as a discipline see: Jensen, J. (2011). "The argument for a disciplinary approach to emergency management higher education." In J. Hubbard, Challenges of Emergency Management in Higher Education, (18-47). Fairfax, VA: Public Entity Risk Institute.

234 For more on the discussion of the professionalization of emergency management see: Wilson, J., and Oyola-Yemaiel, A. (2001). "The evolution of emergency management and the advancement toward a profession in the United States and Florida." *Safety Science* 39(1-2), 117-131.

235 For more on the Red Cross response to the Galveston hurricane and the

early history of the organization see: Hurd, C. (1959). *The Compact History of The American Red Cross.* Hawthorn Books.

236   For more see Butler, D. (2020). "Focusing Events In The Early Twentieth Century: A Hurricane, Two Earthquakes, and A Pandemic." In Rubin, C. (ed). *Emergency Management: The American Experience,* 3rd Ed. Routledge.

237   For more on the different types of organizations that are involved in disaster work see: Webb, G. R. (1999). "Individual and Organizational Response to Natural Disasters and Other Crisis Events: The Continuing Value of the DRC Typology." Preliminary Paper #227. Newark, DE: Disaster Research Center, University of Delaware.

An Unplanned Response

238   For more on spontaneous volunteers read: Twigg, J., and Mosel, I. (2017). "Emergent groups and spontaneous volunteers in urban disaster response." *Environment and Urbanization* 29(2), 443–458.

239   For a brief summary of disaster convergence see: Auf der Heide, E. A. (2003). "Convergence Behavior in Disasters." *Annals of Emergency Medicine* 41(4), 463–466.

240   Kempton, K. G. (2015, January 12). "Donating A Single Rollerblade Is Not Going To Help Disaster Victims." NPR. https://www.npr.org/sections/goatsandsoda/2015/01/12/376716063/donating-a-single-rollerblade-is-not-going-to-help-disaster-victims.

241   Shapiro, A. (2015, December 14). "Bears for Newtown: Tax Assessor Recalls Gift Influx After Sandy Hook Shooting." [Radio]. NPR. https://www.npr.org/2015/12/14/459718247/bears-for-newtown-tax-assessor-recalls-gift-influx-after-sandy-hook-shooting.

242   For a review of the complications unrequested donations can cause see: Neal, D. (1994). "The consequences of excessive unrequested donations: The case of Hurricane Andrew." *Disaster Management* 6(1).

243   N.A. (2017, September 3). "Best intentions: When disaster relief brings anything but relief." CBS News. https://www.cbsnews.com/news/best-intentions-when-disaster-relief-brings-anything-but-relief.

244   N.A. (2017, September 3). "Best intentions: When disaster relief brings anything but relief." CBS News. https://www.cbsnews.com/news/best-intentions-when-disaster-relief-brings-anything-but-relief.

245   For more on the disproportionate impacts experienced by undocumented immigrants and barriers to accessing aid during disaster read: Méndez, M., Flores-Haro, G., and Zucker, L. (2020). "The (in)visible victims of disaster: Understanding the vulnerability of undocumented Latino/a and indigenous immigrants." *Geoforum* 116, 50–62.

Up to Our Bottoms in Alligators

246   For more on the emergence of unplanned coordination in disasters see: Stallings, R. A., and Quarantelli, E. L. (1985). "Emergent Citizen Groups and Emergency Management." *Public Administration Review* 45, 93–100.

247 For a more recent look at volunteer fatigue see: Montano, S. (2020, June 18). "Disaster Fatigue Is Real—and the Coronavirus Could Make It Worse." Gizmodo. https://earther.gizmodo.com/disaster-fatigue-is-real-and-the-coronavirus-could-make-1844079719.

PART FOUR

Damage Control

248 For more on how the impacts of climate change are already being felt in Grand Isle and southeast Louisiana see: Welch, M. P. (2019, November 11). "What It's Like to Live in a City That's Slowly Drowning." *Vice*. https://www.vice.com/en/article/ne8gw7/new-orleans-is-slowly-drowning-thanks-to-climate-change.

Climate Change Is Already Here

249 USGCRP. (2017): Climate Science Special Report: Fourth National Climate Assessment, Volume I [Wuebbles, D.J., Fahey, D.W., Hibbard, K.A., Dokken, D.J., Steward, B.C., and Maycock, T.K. (eds.)]. U.S. Global Change Research Program, Washington, DC, USA, 470pp. https://science2017.globalchange.gov/.

250 USGCRP. (2017): Climate Science Special Report: Fourth National Climate Assessment, Volume I [Wuebbles, D.J., Fahey, D.W., Hibbard, K.A., Dokken, D.J., Steward, B.C., and Maycock, T.K. (eds.)]. U.S. Global Change Research Program, Washington, DC, USA, 470pp. https://science2017.globalchange.gov/.

251 USGCRP. (2017): Climate Science Special Report: Fourth National Climate Assessment, Volume I [Wuebbles, D.J., Fahey, D.W., Hibbard, K.A., Dokken, D.J., Steward, B.C., and Maycock, T.K. (eds.)]. U.S. Global Change Research Program, Washington, DC, USA, 470pp. https://science2017.globalchange.gov/.

252 NASA. (2003). "European Heat Wave." Earth Observatory. https://earthobservatory.nasa.gov/images/3714/european-heat-wave.

253 Khamsi, R. (2004, December 1). "Human activity implicated in Europe's 2003 heat wave." *Nature*. https://www.nature.com/news/2004/041129/full/041129-6.html.

254 For more on the impacts of the 2003 heatwave: United Nations Environment Programme. (2004). "Impacts of summer 2003 heat wave in Europe." Environment Alert Bulletin. https://www.unisdr.org/files/1145_ewheatwave.en.pdf.

255 Mitchell, D., et al. (2016). "Attributing human mortality during extreme heat waves to anthropogenic climate change." *Environmental Research Letters* 11(7).

256 Stone, W. (2018, July 9). "Phoenix Tries To Reverse Its 'Silent Storm' Of Heat Deaths." NPR. https://www.npr.org/2018/07/09/624643780/phoenix-tries-to-reverse-its-silent-storm-of-heat-deaths.

257 Holloway, M. (2019). "As Phoenix Heats Up, the Night Comes Alive." *New*

York Times. https://www.nytimes.com/interactive/2019/climate/phoenix-heat.html.

258 Krol, D. U. (2020, August 27). "In Phoenix, rising temperatures day and night kill more people each year." *Arizona Republic.* https://www.azcentral.com/in-depth/news/2020/08/26/heat-killing-more-people-cities-sizzle-hotter-temperatures/4553439002.

259 Irfan, U. (2019, September 9). "100 degrees for days: the looming Phoenix heat wave that could harm thousands." *Vox.* https://www.vox.com/energy-and-environment/2019/9/9/20804544/climate-change-phoenix-heat-wave-deaths-extreme-weather.

260 Buchele, M. (2020, August 28). "Poor Neighborhoods Feel Brunt Of Rising Heat. Cities Are Mapping Them To Bring Relief." NPR. https://www.npr.org/2020/08/28/905922215/poor-neighborhoods-feel-brunt-of-rising-heat-cities-are-mapping-them-to-bring-re.

261 Cappucci, M. (2020, June 19). "Sizzling heat engulfs northern New England, challenging modern records." *Washington Post.* https://www.washingtonpost.com/weather/2020/06/19/record-heat-maine.

262 NOAA. "1981-2010 Climate Normals." National Centers for Environmental Information. https://www.ncdc.noaa.gov/cdo-web/datatools/normals.

263 Sarnacki, A. (2018, January 5). "20 years later, memories of Maine's Ice Storm of '98 still fresh." *Bangor Daily News.* https://bangordailynews.com/2018/01/05/news/frozen-in-time-memories-of-the-ice-storm-of-98. And Graham, G. (2018, January 4). "Maine's historic ice storm of 1998 brought extraordinary destruction—and cooperation." *Portland Press Herald.* https://www.pressherald.com/2018/01/04/ice-storm-of-1998-brought-extraordinary-destruction-and-cooperation.

264 NOAA. (n.d.). "Climate Change and Extreme Snow in the U.S." National Centers for Environmental Information. https://www.ncdc.noaa.gov/news/climate-change-and-extreme-snow-us.

265 For a review of this event see: Lott, N. (1993). "The Big One! A Review of the March 12-14, 1993 'Storm of the Century.'" NOAA. https://www1.ncdc.noaa.gov/pub/data/techrpts/tr9301/tr9301.pdf.

266 For more on the 2014 snowstorm: Raymond, J. (2020, January 28). "Go outside and appreciate the clear conditions, because six years ago today was Snowmageddon." WXIA. https://www.11alive.com/article/news/history/atlanta-snowmageddon-2014-anniversary/85-bcf15186-4d02-4a47-8025-5a2cc1f3fdac.

267 Harrington, S. (2019, April 2). "Did climate change cause the flooding in the Midwest and Plains?" Yale Climate Connections. https://yaleclimateconnections.org/2019/04/did-climate-change-cause-midwest-flooding.

268 For more on the Midwest 2019 spring floods see: Simpson, A. (2019, April 16). "Midwest farmers suffer after floods: 'I got my life in this ground.'" Pew Trusts. https://www.pewtrusts.org/en/research-and-analysis/blogs/stateline/2019/04/16/midwest-farmers-suffer-after-floods-i-got-my-life-in-this-ground.

269 For more on the 2009 Flood efforts in Fargo see: Davey, M. (2009, March

25). "Fargo Works to Hold Back Rapidly Rising River." *New York Times.* https://www.nytimes.com/2009/03/26/us/26flood.html.

270 For more on flood mitigation efforts in Fargo see: Headwaters Economics (2020). "Building for the Future: Five Midwestern Communities Reduce Flood Risk." https://headwaterseconomics.org/wp-content/uploads/FloodCaseStudies_LowRes.pdf.

271 For more on sandbagging efforts in Fargo see: N.A. (2019, March 26). "Volunteers Start Sandbagging on First Day of Fargo Sandbag Central." KVRR News. https://www.kvrr.com/2019/03/26/volunteers-start-sandbagging-on-first-day-of-fargo-sandbag-central.

272 Donegan, B. (2019, May 22). "2019 Mississippi River Flood the Longest-Lasting Since the Great Flood of 1927 in Multiple Locations." The Weather Channel. https://weather.com/news/weather/news/2019-05-14-one-of-longest-lived-mississippi-river-floods-since-great-flood-1927.

273 Almukhtar, S., Migliozzi, B., Schwartz, J., and Williams, J. (2019, September 11). "The Great Flood of 2019: A Complete Picture of a Slow-Motion Disaster." *New York Times.* https://www.nytimes.com/interactive/2019/09/11/us/midwest-flooding.html.

274 O'Connell, P. M. and Briscoe, T. (2020, January 16). "In 2019—the 2nd wettest year ever in the U.S.—flooding cost Illinois and the Midwest $6.2 billion. Scientists predict more waterlogged days ahead." *Chicago Tribune.* https://www.chicagotribune.com/news/environment/ct-climate-disasters-cost-midwest-20200115-jubchhqe7bfdnolpw3z7cwjwvm-story.html.

275 For more on the 2019 flooding in Nebraska: Gaarder, N. (2020, January 15). "'An unprecedented event': Nebraska's losses from 2019 flooding, blizzard exceed $3.4 billion." *Omaha World-Herald.* https://omaha.com/news/plus/an-unprecedented-event-nebraskas-losses-from-2019-flooding-blizzard-exceed-3-4-billion/article_1bbe1c5c-17de-53f2-a18f-459a1b5a1cdd.html.

276 USGCRP. (2017): Climate Science Special Report: Fourth National Climate Assessment, Volume I [Wuebbles, D.J., Fahey, D.W., Hibbard, K.A., Dokken, D.J., Steward, B.C., and Maycock, T.K. (eds.)]. U.S. Global Change Research Program, Washington, DC, USA, 470pp. https://science2017.globalchange.gov/.

277 Berwyn, B. (2019, October 7). "In the Mountains, Climate Change Is Disrupting Everything, from How Water Flows to When Plants Flower." Inside Climate News. https://insideclimatenews.org/news/07102019/mountain-climate-change-disruption-glaciers-water-ecosystems-agriculture-plants-food.

278 Davenport, C. (2014, May 13). "Climate Change Deemed Growing Security Threat by Military Researchers." *New York Times.* https://www.nytimes.com/2014/05/14/us/politics/climate-change-deemed-growing-security-threat-by-military-researchers.html.

279 USGCRP. (2017): Climate Science Special Report: Fourth National Climate Assessment, Volume I [Wuebbles, D.J., Fahey, D.W., Hibbard, K.A., Dokken, D.J., Steward, B.C., and Maycock, T.K. (eds.)]. U.S. Global Change

Research Program, Washington, DC, USA, 470pp. https://science2017.globalchange.gov/.

280 For more on the factors that are contributing to increasing fires see: Borunda, A. (2020, September 17). "The science connecting wildfires to climate change." *National Geographic.* https://www.nationalgeographic.com/science/2020/09/climate-change-increases-risk-fires-western-us.

281 Perry, D. (2019, October 28). "California's fire season is longer and deadlier than ever, causing annual treks south by Oregon firefighters." *The Oregonian.* https://www.oregonlive.com/environment/2019/10/californias-fire-season-is-longer-and-deadlier-than-ever-causing-annual-treks-south-by-oregon-firefighters.html.

282 Valdez, J. (2019, May 7). "Fire Official: 'It's no longer a fire season, we're now calling it a fire year.'" *The Orange County Register.* https://www.ocregister.com/2019/05/07/official-its-no-longer-a-fire-season-were-calling-it-a-fire-year.

283 Kaplan, S. and Eilperin, J. (2020, September 16). "Trump's plan for managing forests won't save us in a more flammable world, experts say." *Washington Post.* https://www.washingtonpost.com/climate-environment/2020/09/16/fires-climate-change.

For more on these changes read: Henson, B. (2020, September 8). "Autumn could deliver the worst of California's 2020 fire season." Yale Climate Connections. https://yaleclimateconnections.org/2020/09/autumn-could-deliver-the-worst-of-californias-2020-fire-season.

284 Petley, D. (2020, August 11). "Landslides after wildfires." American Geophysical Union. https://blogs.agu.org/landslideblog/2020/08/11/california-wildfire-research-1.

285 Cannon, S. H., and DeGraff, J. (2009). "The Increasing Wildfire and Post-Fire Debris-Flow Threat in Western USA, and Implications for Consequences of Climate Change." In *Landslides–Disaster Risk Reduction,* (177-190). Springer, Berlin, Heidelberg.

286 NASA. (2018). "Deadly Debris Flows in Montecito." Earth Observatory. https://earthobservatory.nasa.gov/images/91573/deadly-debris-flows-in-montecito.

287 Fischer, E. (2020, November 16). "Retreating Glacier Presents Landslide Threat, Tsunami Risk in Alaskan Fjord." NASA. https://www.nasa.gov/feature/esnt/2020/retreating-glacier-presents-landslide-threat-tsunami-risk-in-alaskan-fjord.

288 Palmer, J. (2020, November 23). "A Slippery Slope: Could Climate Change Lead to More Landslides?" *Eos.* https://eos.org/features/a-slippery-slope-could-climate-change-lead-to-more-landslides.

289 Thevenot, B. (2019, July 12). "Storm Barry threatens to overwhelm New Orleans' century-old pump system." Reuters. https://www.reuters.com/article/us-storm-barry-pumps/storm-barry-threatens-to-overwhelm-new-orleans-century-old-pump-system-idUSKCN1U7278.

290 Wuebbles, D. J., Fahey, D. W., and Hibbard, K. A. (2017). Climate Science

Special Report: Fourth National Climate Assessment, Volume I. https://science2017.globalchange.gov/.

291   Wuebbles, D. J., Fahey, D. W., and Hibbard, K. A. (2017). Climate Science Special Report: Fourth National Climate Assessment, Volume I. https://science2017.globalchange.gov/.

292   N.A. (2016, September 7). "Climate change increased chances of record rains in Louisiana by at least 40 percent." NOAA. https://www.noaa.gov/media-release/climate-change-increased-chances-of-record-rains-in-louisiana-by-at-least-40-percent#:~:text=Human%2Dcaused%20climate%20warming%20increased,role%20of%20climate%20on%20the.

293   Montano, S. (2018, May 31). "America Is Flooding, and It's Our Fault." City Lab. https://www.bloomberg.com/news/articles/2018-05-31/why-america-has-to-get-serious-about-disaster-mitigation.

294   Fountain, H. (2017, December 13). "Scientists link Hurricane Harvey's record rainfall to climate change." New York Times. https://www.nytimes.com/2017/12/13/climate/hurricane-harvey-climate-change.html?login=email&auth=login-email.

295   Reed, K. A., Stansfield, A. M., Wehner, M. F., and Zarzycki, C. M. (2020). "Forecasted attribution of the human influence on Hurricane Florence." Science Advances 6(1).

296   Masters, J. (2020, August 27). "Climate change is causing more rapid intensification of Atlantic hurricanes." Yale Climate Connections. https://yaleclimateconnections.org/2020/08/climate-change-is-causing-more-rapid-intensification-of-atlantic-hurricanes.

297   NOAA. (2020, September 23). "Global Warming and Hurricanes." Geophysical Fluid Dynamics Laboratory. https://www.gfdl.noaa.gov/global-warming-and-hurricanes.

298   Freedman, A. (2012, November 1). "How Global Warming Made Hurricane Sandy Worse." Climate Central. https://www.climatecentral.org/news/how-global-warming-made-hurricane-sandy-worse-15190#:~:text=Katharine%20Hayhoe%2C%20a%20climate%20researcher,sinking%20land%2C%20and%20ocean%20currents.

299   Samenow, J. (2012, November 15). "Climate change did not cause Superstorm Sandy, but probably intensified its effects." Washington Post. https://www.washingtonpost.com/blogs/capital-weather-gang/post/the-whole-truth-about-superstorm-sandy-and-climate-change/2012/11/15/d3b7ceea-29e4-11e2-bab2-eda299503684_blog.html.

300   Flavelle, C. (2020, July 14). "New Data Shows an 'Extraordinary' Rise in U.S. Coastal Flooding." New York Times. https://www.nytimes.com/2020/07/14/climate/coastal-flooding-noaa.html.

301   Sweet, W. W. V., et al. (2020). "2019 State of US High Tide Flooding with a 2020 Outlook." NOAA Technical Report. https://tidesandcurrents.noaa.gov/publications/Techrpt_092_2019_State_of_US_High_Tide_Flooding_with_a_2020_Outlook_30June2020.pdf.

302   Davenport, C. and Robertson, C. (2016, May 2). "Resettling the First

American 'Climate Refugees.'" *New York Times*. https://www.nytimes.com/2016/05/03/us/resettling-the-first-american-climate-refugees.html.

303   The term *refugee* can have complicated connotations in the disaster context. The term *refugee* is used to describe someone who is forced to leave their country because of persecution, war, or violence. People who are displaced during a disaster within their own country may be referred to as "an evacuee" or an "internally displaced person." Distinction in these terms has important legal implications. Further, there was widespread controversy over the misuse of the term *refugee* used to describe New Orleanians displaced following Katrina. More can be read here: Masquelier, A. (2006). "Why Katrina's Victims Aren't Refugees: Musings on a 'Dirty' Word." *American Anthropologist* 108(4), 735-743.

304   For more on flooding in Venice read: Keahey, J. (2002). *Venice Against The Sea: A City Besieged*. Thomas Dunne Books.

305   Bjarnason, E. (2018, January 16). "Why Iceland is Turning Purple." *Hakai Magazine*. https://www.hakaimagazine.com/features/why-iceland-is-turning-purple.

306   Liska, I. (2015). "Managing an International River Basin Toward Water Quality Protection: The Danube Case." In *The Danube River Basin* (1-19). Springer, Berlin, Heidelberg. https://www.icpdr.org/main/sites/default/files/nodes/documents/1stdfrmp-final.pdf.

307   Cappucci, M. and Daileda, C. (2019, August 9). "Kerala floods turn deadly as India suffers relentless monsoonal deluge." *Washington Post*. https://www.washingtonpost.com/weather/2019/08/09/kerala-floods-turn-deadly-india-suffers-horrific-monsoonal-deluge.

308   Ortiz, F. (2017, September 8). "Cartagena struggles to get pioneering climate plan into action." Reuters. https://www.reuters.com/article/us-colombia-climatechange-cartagena/cartagena-struggles-to-get-pioneering-climate-plan-into-action-idUSKCN1BK00N.

309   For more on this emerging science see: Otto, F. (2020). *Angry Weather: Heat Waves, Floods, Storms, and the New Science of Climate Change*. Greystone Books.

310   Pidcock, R., Pearce, R., and McSweeney, R. (2020, April 15). "Mapped: How climate change affects extreme weather around the world." Carbon Brief. https://www.carbonbrief.org/mapped-how-climate-change-affects-extreme-weather-around-the-world.

Facing the Consequences

311   USGCRP. (2017): Climate Science Special Report: Fourth National Climate Assessment, Volume I [Wuebbles, D.J., Fahey, D.W., Hibbard, K.A., Dokken, D.J., Steward, B.C., and Maycock, T.K. (eds.)]. U.S. Global Change Research Program, Washington, DC, USA, 470pp. https://science2017.globalchange.gov/.

312   Wallace-Wells, D. (2017). "The Uninhabitable Earth." *New York Magazine*. http://nymag.com/daily/intelligencer/2017/07/climate-change-earth-too-hot-for-humans.html.

313 Berwyn, B. (2020, August 3). "The Worst-Case Scenario for Global Warming Tracks Closely With Actual Emissions." *Inside Climate News*. https://insideclimatenews.org/news/03082020/climate-change-scenarios-emissions.

314 IPCC (2018). Global Warming of 1.5°C. IPCC Special Report. https://www.ipcc.ch/sr15.

315 Hayhoe, K., et al. (2017). "Climate Models, Scenarios, and Projections." In *Climate Science Special Report: Fourth National Climate Assessment*, Volume I. U.S. Global Change Research Program. https://science2017.globalchange.gov/chapter/4.

316 Montano, S. (2019, September 19). "What's missing from the Democrats' climate conversation so far." *The Week*. https://theweek.com/articles/863831/whats-missing-from-democrats-climate-conversation-far.

317 Montano, S. (2019, September 6). "The Climate Conversation Must Include Emergency Management." *Disasterology*. http://www.disaster-ology.com/blog/2019/9/6/the-climate-conversation-must-include-emergency-management.

318 Savitt, A. and Montano, S. (2019, October 14). "Evaluating the Democratic Candidates' Disaster Policy." *Disasterology*. http://www.disaster-ology.com/blog/2019/10/14/14komh82s5zcx9ul08b0h69nm2nefy.

"A Real Catastrophe Like Katrina"

319 For more on Puerto Rico and Hurricane Maria read: Bonilla, Y., and Le-Brón, M. (Eds.). (2019). *Aftershocks of Disaster: Puerto Rico Before And After The Storm*. Haymarket Books. And Klein, N. (2018). *The Battle For Paradise: Puerto Rico Takes On The Disaster Capitalists*. Haymarket Books.

320 See this special issue for more: Rivera, F. I. (2020). "Puerto Rico's population before and after Hurricane Maria." *Population and Environment*. https://www.ncbi.nlm.nih.gov/pmc/articles/PMC7387120.

321 Wilson, C. (2017, March 17). "How Much Does Puerto Rico Owe? 4 Charts That Put Puerto Rican Debt in Perspective." *Money Morning*. https://moneymorning.com/2017/03/17/how-much-does-puerto-rico-owe-4-charts-that-put-puerto-rican-debt-in-perspective. And Sullivan, L. (2018, May 2). "How Puerto Rico's Debt Created A Perfect Storm Before The Storm." NPR. https://www.npr.org/2018/05/02/607032585/how-puerto-ricos-debt-created-a-perfect-storm-before-the-storm.

322 Foxman, S. (2013, September 19). "Puerto Rico is living an impoverished debt nightmare reminiscent of southern Europe or Detroit." *Quartz*. https://qz.com/125654/puerto-rico-is-living-an-impoverished-debt-nightmare-reminiscent-of-southern-europe-or-detroit.

323 Long-Garcia, J. (2017, December 27). "What the hurricanes revealed about Puerto Rico." *America: The Jesuit Review*. https://www.americamagazine.org/politics-society/2017/12/27/what-hurricanes-revealed-about-puerto-rico.

324 For extensive discussion on vulnerability see: Tierney, K. (2014). *The Social*

Roots of Risk: Producing Disasters, Promoting Resilience. Stanford University Press.

325 Hsiang, S. and Houser, T. (2017, November 29). "Don't Let Puerto Rico Fall Into an Economic Abyss." *New York Times.* https://www.nytimes.com/2017/09/29/opinion/puerto-rico-hurricane-maria.html?_r=1.

326 Hersher, R. (2019, April 17). "Climate Change Was The Engine That Powered Hurricane Maria's Devastating Rains." NPR. https://www.npr.org/2019/04/17/714098828/climate-change-was-the-engine-that-powered-hurricane-marias-devastating-rains.

327 Bessette-Kirton, E. K., et al. (2019). "Landslides Triggered by Hurricane Maria: Assessment of an Extreme Event in Puerto Rico." *GSA Today* 29(6), 4-10.

328 Keellings, D., and Hernández Ayala, J. J. (2019). "Extreme rainfall associated with Hurricane Maria over Puerto Rico and its connections to climate variability and change." *Geophysical Research Letters* 46(5), 2964-2973.

329 For more on Hurricane Irma see: McSweeney, R. and Timperley, J. (2017, December 9). "Media reaction: Hurricane Irma and climate change." *Carbon Brief.* https://www.carbonbrief.org/media-reaction-hurricane-irma-climate-change.

330 Schmidt, S. and Hernández, A. R. (2017, October 1). "Trapped in the mountains, Puerto Ricans don't see help, or a way out." *Washington Post.* https://www.washingtonpost.com/national/trapped-in-the-mountains-puerto-ricans-dont-see-help-or-a-way-out/2017/10/01/7621867e-a647-11e7-ade1-76d061d56efa_story.html?utm_term=.81c3d4ba1b29.

331 Coto, D. (2020, July 24). "Thousands in Puerto Rico still without housing since Maria." *Washington Post.* https://www.washingtonpost.com/world/the_americas/thousands-in-puerto-rico-still-without-housing-since-maria/2020/07/24/9f79ee6c-cd63-11ea-99b0-8426e26d203b_story.html.

332 Campbell, A. F. (2018, August 15). "It took 11 months to restore power to Puerto Rico after Hurricane Maria. A similar crisis could happen again." *Vox.* https://www.vox.com/identities/2018/8/15/17692414/puerto-rico-power-electricity-restored-hurricane-maria.

333 Agence France-Presse, "Trump bashes Puerto Rico's leaders: 'They want everything to be done for them,'" *PRI The World,* September 30, 2017, https://www.pri.org/stories/2017-09-30/trump-bashes-puerto-ricos-leaders-they-want-everything-be-done-them.

334 Mehta, D. (2017, September 28). "The Media Really Has Neglected Puerto Rico." *FiveThirtyEight.* https://fivethirtyeight.com/features/the-media-really-has-neglected-puerto-rico.

335 Shah, A. K. (2017, November 27). "The mainstream media didn't care about Puerto Rico until it became a Trump story." *Washington Post.* https://www.washingtonpost.com/news/posteverything/wp/2017/11/27/the-mainstream-media-didnt-care-about-puerto-rico-until-it-became-a-trump-story/?utm_term=.26e9061bc9c5.

336 For an example of the importance of media coverage and disasters see: Syl-

vester, J. (2008). *The Media and Hurricanes Katrina and Rita: Lost and Found.* Springer.

337   Federal Emergency Management Agency. (2017, November 7). "Hurricane Maria update." Department of Homeland Security. https://www.dhs.gov/news/2017/11/07/hurricane-maria-update.

338   Reardon, M. (2017, September 3). "How the wireless carriers fared during Hurricane Harvey." CNET. https://www.cnet.com/news/hurricane-harvey-phone-service.

339   Stevens, T. (2017, November 22). "Local reporters in Puerto Rico 'forge ahead' through wreckage." *Columbia Journalism Review.* https://www.cjr.org/united_states_project/puerto-rico-journalists-hurricane-maria.php.

340   For more on disaster reporting see: Steffens, M. (2012). *Reporting Disaster on Deadline: A Handbook for Students and Professionals.* Routledge. For a review of the research see: Scanlon, J. (2011). "Research About The Mass Media and Disaster: Never (well hardly ever) The Twain Shall Meet." Journalism Theory and Practice: 233-269. https://training.fema.gov/emiweb/downloads/scanlonjournalism.pdf.

341   Rooney, P. (2018, September 28). "US generosity after disasters: 4 questions answered." *The Conversation.* https://theconversation.com/us-generosity-after-disasters-4-questions-answered-103215.

342   For more on donor fatigue in 2017 see: Penta, S. (2017, November 14). "Why Puerto Rico is getting the brunt of 'donor fatigue.'" *The Conversation.* https://theconversation.com/why-puerto-rico-is-getting-the-brunt-of-donor-fatigue-86036.

343   Sherman, C. (2017, October 8). "Puerto Rico has received millions less in donations than the mainland." *Vice News.* https://www.vice.com/en/article/4347bp/compared-to-harvey-very-few-companies-have-given-toward-maria-relief.

344   Montano, S. (2017). "A foundation for factors that explain volunteer engagement in response and recovery: The case of flooding in East Texas 2016." Doctoral Dissertation. North Dakota State University, North Dakota, USA.

345   Cronk, T. M. (2017, September 22). "Military Officials Outline Hurricane Relief Efforts." US Department of Defense. https://www.defense.gov/Explore/News/Article/Article/1321860/military-officials-outline-hurricane-relief-efforts.

346   Gillette, C. (2017, September 24). "Aid begins to flow to hurricane-hit Puerto Rico." AP News. https://apnews.com/article/06f5077aff384e508e2f2324dae4eb2e.

347   Benen, S. (2017, September 29). "Trump: Puerto Rico is 'an island surrounded by ... big water.'" MSNBC. https://www.msnbc.com/rachel-maddow-show/trump-puerto-rico-island-surrounded-big-water-msna1024996.

348   Chappel, B. (2017, September 13). "Power Outages Persist For Millions In Florida, Georgia And Carolinas After Irma." NPR. https://www.npr.org/sections/thetwo-way/2017/09/13/550674848/power-outages-persist-for-millions-in-florida-georgia-and-carolinas-after-irma.

349 For a timeline of the response to Maria see: Meyer, R. (2017, October 4). "What's Happening With the Relief Effort in Puerto Rico?" *The Atlantic*. https://www.theatlantic.com/science/archive/2017/10/what-happened-in-puerto-rico-a-timeline-of-hurricane-maria/541956.

350 Campbell, A. F. (2017, October 3). "Trump to Puerto Rico: Your hurricane isn't a 'real catastrophe' like Katrina." *Vox*. https://www.vox.com/2017/10/3/16411488/trump-remarks-puerto-rico.

351 Kishore, N. M. et al. (2018, July 12). "Mortality in Puerto Rico after Hurricane Maria." *New England Journal of Medicine*. https://www.nejm.org/doi/full/10.1056/NEJMsa1803972.

352 Beauchamp, Z. (2018, September 13). "Trump's Puerto Rico tweets are the purest expression of his presidency." *Vox*. https://www.vox.com/2018/9/13/17855268/trump-hurricane-maria-puerto-rico-tweets.

353 Florido, A. (2018, June 1). "An Impromptu Memorial To Demand That Puerto Rico's Hurricane Dead Be Counted." NPR. https://www.npr.org/2018/06/01/616216225/an-impromptu-memorial-to-demand-puerto-ricos-hurricane-dead-be-counted.

354 Bourque, L. B., Siegel, J. M., Kano, M., and Wood, M. M. (2007). "Morbidity and Mortality Associated with Disasters." In *Handbook of Disaster Research*, (97-112). Springer, New York, NY.

355 National Academies of Sciences, Engineering, and Medicine. (2020). "A framework for assessing mortality and morbidity after large-scale disasters." Washington, DC: The National Academies Press. https://www.nationalacademies.org/our-work/best-practices-in-assessing-mortality-and-significant-morbidity-following-large-scale-disasters.

356 Goss, K. C. (2016, November 23). "Importance of Presidential Leadership in Emergency Management." Domestic Preparedness. https://www.domesticpreparedness.com/commentary/importance-of-presidential-leadership-in-emergency-management.

357 Siddiqui, F. and Soong, K. (2017, November 11). "Puerto Rico Unity March draws demonstrators in rally for disaster aid." *Washington Post*. https://www.washingtonpost.com/news/local/wp/2017/11/19/puerto-rico-unity-march-draws-thousands-to-mall-in-plea-for-disaster-aid.

358 Hui, M. (2017, October 3). "Oxfam slams the Trump administration for its 'slow and inadequate' response in Puerto Rico." *Washington Post*. https://www.washingtonpost.com/news/worldviews/wp/2017/10/03/oxfam-slams-the-trump-administration-for-its-slow-and-inadequate-response-in-puerto-rico.

359 Willison, C. E., Singer, P. M., Creary, M. S., and Greer, S. L. (2019). "Quantifying inequities in US federal response to hurricane disaster in Texas and Florida compared with Puerto Rico." *BMJ Global Health* 4(1).

360 N.A. (2020, September 25). "FEMA Mismanaged the Commodity Distribution Process in Response to Hurricanes Irma and Maria." Office of Inspector General. https://www.oig.dhs.gov/sites/default/files/assets/2020-09/OIG-20-76-Sep20.pdf.

361 Lopez, G. (2017, October 2). "Is Las Vegas the worst mass shooting in US history? It's surprisingly complicated." *Vox*. https://www.vox.com/identities/2017/10/2/16401510/las-vegas-shooting-deadliest.

362 Federal Emergency Management Agency. (2017, December 29). "FEMA Reflects On Historic Year." Federal Emergency Management Agency. https://www.fema.gov/press-release/20210318/fema-reflects-historic-year.

363 Henry, D. (2017, October 31). "FEMA chief: Feds spending $200 million a day on hurricane, wildfire recovery." *The Hill*. http://thehill.com/policy/energy-environment/358013-fema-chief-feds-spending-200-million-a-day-on-hurricane-wildfire.

364 National Oceanic Atmospheric Administration. (2018, April 26). "Billion Dollar Weather and Climate Disasters: Overview." National Oceanic Atmospheric Administration. https://www.ncdc.noaa.gov/billions.

365 Flatley, D. (2017, November 30). "FEMA's Staff 'Tapped Out' After Mounting Disasters, Leader Says." *Bloomberg*. https://www.bloomberg.com/news/articles/2017-11-30/fema-s-staff-tapped-out-after-mounting-disasters-leader-says.

366 N.A. (2020, September 25). "FEMA Mismanaged the Commodity Distribution Process in Response to Hurricanes Irma and Maria." Office of Inspector General. https://www.oig.dhs.gov/sites/default/files/assets/2020-09/OIG-20-76-Sep20.pdf.

PART FIVE

Preparedness: Anticipating Disaster

367 Talmazan, Y. (2020, February 3). "China's coronavirus hospital built in 10 days opens its doors, state media says." NBC News. https://www.nbcnews.com/news/world/china-s-coronavirus-hospital-built-10-days-opens-its-doors-n1128531.

368 For more on UNMC's experience during the pandemic see: Yong, E. (2020, November 20). "Hospitals Know What's Coming." *The Atlantic*. https://www.theatlantic.com/health/archive/2020/11/americas-best-prepared-hospital-nearly-overwhelmed/617156.

369 Anderson, J. (2020, February 7). "57 Americans land in Omaha to be quarantined, watched for signs of coronavirus." *Omaha World-Herald*. https://www.omaha.com/livewellnebraska/57-americans-land-in-omaha-to-be-quarantined-watched-for-signs-of-coronavirus/article_1b959a4e-254a-5142-af58-4599ee3c5d71.html.

370 Wilson, T. (2020, February 25). "Second person with positive coronavirus test coming to UNMC/Nebraska Medicine." University of Nebraska Medical Center. https://www.unmc.edu/news.cfm?match=25151.

371 Federal Emergency Management Agency. (2019). "National Response Framework." FEMA. https://www.fema.gov/emergency-managers/national-preparedness/frameworks/response.

372 Holshue, M. L., et al. (2020, March 5). "First case of 2019 novel coronavirus

in the United States." *New England Journal of Medicine.* https://www.nejm.org/doi/full/10.1056/NEJMoa2001191.

373 For the full story see: Fink, S. and Baker, M. (2020, March 10). "'It's Just Everywhere Already': How Delays in Testing Set Back the U.S. Coronavirus Response." *New York Times.* https://www.nytimes.com/2020/03/10/us/coronavirus-testing-delays.html.

374 Shiff, B. and Katersky, A. (2017, September 7). "Hurricane Irma hit Barbuda like a 'bomb', prime minister says." ABC News. https://abcnews.go.com/International/hurricane-irma-hit-barbuda-bomb-prime-minister/story?id=49665358.

375 N.A. (2017, September 22). "Hurricane Irma erased 'footprints of an entire civilization' on Barbuda, Prime Minister tells UN." UN News. https://news.un.org/en/story/2017/09/566372-hurricane-irma-erased-footprints-entire-civilization-barbuda-prime-minister.

Surviving on Checklists & Go-Bags

376 Miller, L. (2019, July 17). "Men Know It's Better to Carry Nothing." *The Cut.* https://www.thecut.com/2019/07/if-men-carried-purses-would-they-clean-up-messes.html.

377 N.A. (2019). Mobile Fact Sheet. Pew Research Center. *https://www.pewresearch.org/internet/fact-sheet/mobile.*

378 FEMA. (2020). "2020 NHS Data Digest: Summary Results." Federal Emergency Management Agency. https://community.fema.gov/story/2020-NHS-Data-Digest-Summary-Results.

379 For a critique of this research see: Nojang, E. N., and Jensen, J. (2020). "Conceptualizing Individual and Household Disaster Preparedness: The Perspective from Cameroon." *International Journal of Disaster Risk Science,* 1-14.

380 For an investigation on the COVID testing failures see: Boburg, S., O'Harrow, R., Satija, N., and Goldstein, A. (2020, April 3). "Inside the coronavirus testing failure: Alarm and dismay among the scientists who sought to help." *Washington Post.* https://www.washingtonpost.com/investigations/2020/04/03/coronavirus-cdc-test-kits-public-health-labs/?arc404=true.

381 Wieczner, J. (2020, May 18). "The case of the missing toilet paper: How the coronavirus exposed U.S. supply chain flaws." *Fortune.* https://fortune.com/2020/05/18/toilet-paper-sales-surge-shortage-coronavirus-pandemic-supply-chain-cpg-panic-buying.

382 For a timeline of the COVID response see: Taylor, D. B. (2021, January 10). "A Timeline of The Coronavirus Pandemic." *New York Times.* https://www.nytimes.com/article/coronavirus-timeline.html.

383 Schneider, S. K. (2018). "Government Response to Disasters: Key Attributes, Expectations, and Implications." In Rodríguez, H., Donner, W., and Trainor, J. E. eds. *The Handbook of Disaster Research,* (551-568). Springer.

384 McMinn, S. (2020, May 1). "Mobile Phone Data Show More Americans Are Leaving Their Homes, Despite Orders." NPR. https://www.npr.

org/2020/05/01/849161820/mobile-phone-data-show-more-americans-are-leaving-their-homes-despite-orders.

385 To see the full model and see the research: Nojang, E. N., and Jensen, J. (2020). "Conceptualizing Individual and Household Disaster Preparedness: The Perspective from Cameroon." *International Journal of Disaster Risk Science*, 1-14.

Designing an Emergency Management System

386 For an overview of the history of emergency management see: Rubin, C. B. (Ed.). (2019). *Emergency Management: The American Experience.* Routledge.

387 Unless otherwise noted the history of the Mississippi River flood here comes from: Barry, J. M. (2007). *Rising Tide: The Great Mississippi Flood of 1927 and How It Changed America.* Simon & Schuster.

388 Rubin, C. B. (Ed.). (2019). *Emergency Management: The American Experience.* Routledge.

389 Mizelle Jr., R. M. (2014). *Backwater Blues: The Mississippi Flood of 1927 in the African American Imagination.* University of Minnesota Press.

390 Whyte, K. (2017). *Hoover: An Extraordinary Life in Extraordinary Times.* Vintage.

391 Rubin, C. B. (Ed.). (2019). *Emergency Management: The American Experience.* Routledge.

392 For an extensive history of the civil defense era see: Yoshpe, H. B. (1981). *"Our Missing Shield: The US Civil Defense Program in Historical Perspective."* Federal Emergency Management Agency. https://fas.org/irp/agency/dhs/fema/civildef-1981.pdf.

393 For an example of how these guidances manifested locally see: Knowles, S. G. (2007). "Defending Philadelphia: A historical case study of civil defense in the early Cold War." *Public Works Management & Policy* 11(3), 217-232.

394 Provencio, A. L. (2017). *Gender and Representative Bureaucracy: Opportunities and Barriers in Local Emergency Management Agencies.* (Doctoral dissertation). Oklahoma State University.

395 National Governors' Association (NGA) (1978). Emergency Preparedness Project and Comprehensive Emergency Management: A Governor's Guide. Washington, DC.

396 May, P. J. (1985). "FEMA's Role in Emergency Management: Examining Recent Experience." *Public Administration Review* 45, 40-48.

397 Sylves, R. T. (2019). *Disaster Policy and Politics: Emergency Management and Homeland Security.* CQ Press.

398 Roberts, P. (2005). "The Master of Disaster as Bureaucratic Entrepreneur." *PS, Political Science & Politics* 38(2), 331.

399 Holdeman, E. and Patton, A. (2008, December 12). "Project Impact Initiative to Create Disaster-Resistant Communities Demonstrates Worth in Kansas Years Later." Emergency Management. https://www.govtech.com/em/disaster/Project-Impact-Initiative-to.html.

400 N.A. (2001). "Road Map for National Security: Imperative for Change." The United States Commission on National Security/21st Century. https://www.hsdl.org/?abstract&did=2079.

401 Perrow, C. (2005). "Using Organizations: The Case of FEMA." *Homeland Security Affairs* 4(1). https://www.hsaj.org/articles/687?utm_source=rss&utm_medium=rss&utm_campaign=using-organizations-the-case-of-fema#fn10.

402 Adamski, T., Kline, B., and Tyrrell, T. (2006). "FEMA Reorganization and the Response to Hurricane Disaster Relief." *Perspectives in Public Affairs* 3, 3-36.

403 Sylves, R. T. (2019). *Disaster Policy and Politics: Emergency Management and Homeland Security.* CQ Press.

404 Kean, T., and Hamilton, L. (2004). "The 9/11 commission report: Final report of the national commission on terrorist attacks upon the United States" (Vol. 3). Government Printing Office.

405 Tierney, K. (2005). "The 9/11 Commission and Disaster Management: Little Depth, Less Context, Not Much Guidance." *Contemporary Sociology* 34(2), 115-120.

406 Kendra, J. M., and Wachtendorf, T. (2016). *American Dunkirk: The Waterborne Evacuation of Manhattan on 9/11.* Temple University Press.

407 Tierney, K. J. (2001, November 1). "Strength of a City: A Disaster Research Perspective on the World Trade Center Attack." Items: Insights from the Social Sciences. https://items.ssrc.org/after-september-11/strength-of-a-city-a-disaster-research-perspective-on-the-world-trade-center-attack/.

408 Tierney, K. J. (2001, November 1). "Strength of a City: A Disaster Research Perspective on the World Trade Center Attack." Items: Insights from the Social Sciences. https://items.ssrc.org/after-september-11/strength-of-a-city-a-disaster-research-perspective-on-the-world-trade-center-attack/.

409 Sylves, R. T. (2019). *Disaster Policy and Politics: Emergency Management and Homeland Security.* CQ Press.

410 For more on how post-9/11 changes were implemented across the country see: Jensen, J. A. (2010). *Emergency Management Policy: Predicting National Incident Management System (NIMS) Implementation Behavior.* (Doctoral Dissertation). Fargo, ND: North Dakota State University.

411 For further analysis read: Waugh Jr., W. L. (2006). "The Political Costs of Failure in the Katrina and Rita Disasters." *The ANNALS of the American Academy of Political and Social Science* 604(1), 10-25.

412 Waugh Jr., W. L., and Sylves, R. T. (2002). "Organizing the War on Terrorism." *Public Administration Review* 62, 145-153.

413 Quoted in Tierney, K. J. (2007). "Recent developments in US homeland security policies and their implications for the management of extreme events." In *Handbook of Disaster Research*, (405-412). Springer, New York, NY.

414 Birkland, T. A. (2009). "Disasters, Catastrophes, and Policy Failure in the Homeland Security Era 1." *Review of Policy Research* 26(4), 423-438.

415   Moynihan, D. P. (2009). "The Response to Hurricane Katrina." *International Risk Governance Council*, 1–11.

416   Neuman, J. (2006, February 2). "Report Blames Katrina Response on Chertoff." *Lost Angeles Times*. https://www.latimes.com/archives/la-xpm-2006-feb-02-na-katrina2-story.html.

417   Townsend, F. F. (2006). "The Federal Response to Hurricane Katrina: Lessons Learned." Washington, DC: The White House.

418   Tierney, K. J. (2007, July 31). Testimony by Kathleen J. Tierney. House Committee on Oversight and Government Reform. https://www.globalsecurity.org/security/library/congress/2007_h/070731-tierney.pdf.

419   IAEM. (2008, November 19). "IAEM-USA Calls for Restoring FEMA to Independent Agency Status." News Release. http://www.iaem.com/documents/IAEM-USACallsForFEMAtoBeRestoredtoanIndependent-Agency.pdf.

420   Lombardi, K. (2005, August 30). "Hillary Says FEMA Should Be Set Free." *The Village Voice*. https://www.villagevoice.com/2005/08/30/hillary-says-fema-should-be-set-free.

421   N.A. (2009, May 14). "An Independent FEMA: Restoring the Nation's Capabilities for Effective Emergency Management and Disaster Response." Hearing before the Committee on Transportation and Infrastructure. https://www.hsdl.org/?abstract&did=36763.

422   Cigler, B. A. (2009). "Emergency Management Challenges for the Obama Presidency." *International Journal of Public Administration* 32(9), 759–766.

Planning for Walruses

423   Clarke, L. (1999). *Mission Improbable: Using Fantasy Documents to Tame Disaster*. University of Chicago Press.

424   AP. (2010, June 10). "BP spill response plans severely flawed: Government-approved document listed dead scientist as expert, walruses as gulf inhabitants." AP. https://www.mlive.com/politics/2010/06/bp_spill_response_plans_severe.html.

425   Noguchi, Y. (2010, June 4). "Helping The Pros: Amateur Ideas To Stop The Oil Spill." NPR. https://www.npr.org/templates/story/story.php?storyId=127481460.

426   Keegan, R. (2010, June 3). "James Cameron's Oil-Spill Brainstorming Session: 'It Was Time to Sound the Horn.'" *Vanity Fair*. https://www.vanityfair.com/hollywood/2010/06/james-camerons-oil-spill-brainstorming-session-it-was-time-to-sound-the-horn.

427   Obama, B. (2020). *A Promised Land*. Crown.

428   Baltimore, C. (n.d.). "Obama's oil spill chief reflects a year later." Reuters. https://www.reuters.com/article/instant-article/idUSTRE73J3XO20110420.

429   Birkland, T. A., and DeYoung, S. E. (2011). "Emergency Response, Doctrinal Confusion, and Federalism in the Deepwater Horizon Oil Spill." *Publius: The Journal of Federalism* 41(3), 471–493.

430   Labadie, J. R. (1984). "Problems in Local Emergency Management." *Environmental Management* 8(6), 489-494.

431   Neal, D. M., and Phillips, B. D. (1995). "Effective Emergency management: Reconsidering the Bureaucratic Approach." *Disasters* 19(4), 327-337.

432   Savitt, A. M. (2020). "The Role of the County Emergency Manager in Disaster Mitigation." (Doctoral dissertation.) North Dakota State University.

433   Jensen, J., Bundy, S., Thomas, B., and Yakubu, M. (2014). "The County Emergency Manager's Role in Recovery." *International Journal of Mass Emergencies & Disasters* 32(1).

434   Darby, L. (2020, June 1). "This Is How Much Major Cities Prioritize Police Spending Versus Everything Else." *GQ.* https://www.gq.com/story/cops-cost-billions.

435   New York City Emergency Management (2018). "Report of the Finance Division on the Fiscal 2019 Preliminary Budget and the Fiscal 2018 Preliminary Mayor's Management Report." The Council of the City of New York. https://council.nyc.gov/budget/wp-content/uploads/sites/54/2018/03/FY19-New-York-City-Emergency-Management.pdf.

436   McEntire, D. A. (2018). "Local Emergency Management Organizations." In Rodríguez, H., Donner, W., and Trainor, J. E. eds. *The Handbook of Disaster Research.* Springer, (168-182).

437   McEntire, D. A., and Myers, A. (2004). "Preparing Communities for Disasters: Issues and Processes for Government Readiness." *Disaster Prevention and Management* 13(2), 140-152.

438   Alexander, E. (2009). "Dilemmas in Evaluating Planning, or Back to Basics: What is Planning For?" *Planning Theory & Practice* 10(2), 233-244.

439   Perry, R. W., Lindell, M. K., and Tierney, K. J. (Eds.). (2001). *Facing the Unexpected: Disaster Preparedness and Response in the United States.* Joseph Henry Press.

440   Brooks, M. (2002). *Planning Theory for Practitioners.* Chicago: Planners Press, American Planning Association.

441   Rubin, C. B. (2009). "Long Term Recovery From Disasters—The Neglected Component of Emergency Management." *Journal of Homeland Security and Emergency Management* 6(1).

442   Alexander, E. (2009). "Dilemmas in Evaluating Planning, or Back to Basics: What Is Planning For?" *Planning Theory & Practice* 10(2), 233-244.

443   Somers, S., and Svara, J. H. (2009). "Assessing and Managing Environmental Risk: Connecting Local Government Management with Emergency Management." *Public Administration Review* 69(2), 181-193.

444   For more on vulnerability see: Cutter, S. L., Boruff, B. J., and Shirley, W. L. (2003). "Social Vulnerability to Environmental Hazards." *Social Science Quarterly* 84(2), 242-261.

445   N.A. (2020, July 24). "Historical Flood Risk and Costs." Federal Emergency Management Agency. https://www.fema.gov/data-visualization/historical-flood-risk-and-costs.

446 Roberts, P. S. (2017, September 13). "Why presidents should stop showing up at disaster sites." *Washington Post*. https://www.washingtonpost.com/
news/made-by-history/wp/2017/09/13/why-presidents-should-stop-
showing-up-at-disaster-sites.

447 Crisp, E. and Sentell, W. (2016, August 19). "Donald Trump, Mike Pence
meet Louisiana flood victims, tour hard-hit Baton Rouge neighborhoods."
The Advocate. https://www.theadvocate.com/louisiana_flood_2016/article_
c30c8178-661a-11e6-81c9-3f4b02af0c03.html.

448 Malone, M. A. (2018). "Trump's FEMA Administrator Appointment: The
Pick He Got Correct." *Journal of Public Affairs* 18(2), e1692.

449 N.A. (2017, January 10). "Nomination of General John F. Kelly, USMC
(Ret.), to be Secretary, U.S. Department of Homeland Security" [Video].
U.S. Senate Committee on Homeland Security & Government Affairs.
https://www.hsgac.senate.gov/hearings/nomination-of-general-john-f-
kelly-usmc-ret-to-be-secretary-us-department-of-homeland-security.

450 N.A. (2017, February 15). "Letter Opposing Nomination of Scott Pruitt
as EPA Administrator from 234 Organizations." https://www.nrdc.org/
resources/letter-opposing-nomination-scott-pruitt-epa-administrator-
173-organizations.

451 DelReal, J. A. (2017, March 2). "Ben Carson, outsider with no government experience, confirmed to lead HUD." *Washington Post*. https://www.
washingtonpost.com/politics/ben-carson-outsider-with-no-government-
experience-confirmed-to-lead-hud/2017/03/02/326e5e8e-e8d3-11e6-
80c2-30e57e57e05d_story.html.

452 Scott, D. (2017, September 29). "Tom Price, Trump's scandal-plagued
HHS secretary, is stepping down." *Vox*. https://www.vox.com/policy-and-
politics/2017/9/29/16376220/tom-price-resigns-hhs-secretary.

453 Fandos, N. (2017, March 9). "Trump Weighs Cuts to Coast Guard, T.S.A.
and FEMA to Bolster Border Plan." *New York Times*. https://www.nytimes.
com/2017/03/09/us/politics/trump-budget-coast-guard.html.

454 Lamothe, D., Halsey, A., and Rein, L. (2017, March 7). "To fund border
wall, Trump administration weighs cuts to Coast Guard, airport security."
Washington Post. https://www.washingtonpost.com/world/national-security/
to-fund-border-wall-trump-administration-weighs-cuts-to-coast-guard-
airport-security/2017/03/07/ba4a8e5c-036f-11e7-ad5b-d22680e18d10_
story.html?utm_term=.a7f74eea5d5f.

455 Parlapiano, A. and Aisch, G. (2016, March 16). "Who Wins and Loses in
Trump's Proposed Budget." *New York Times*. https://www.nytimes.com/
interactive/2017/03/15/us/politics/trump-budget-proposal.html.

456 For more analysis of the Trump administration's first year of disaster re-
sponse see: Sylves, R. T. (2019). "Hurricanes Harvey, Irma, and Maria: U.S.
Disaster Management Challenged." In *Disaster Policy and Politics: Emergency
Management and Homeland Security*, (351-388). CQ Press.

457 Pramuk, J. (2017, August 25). "Trump to Texas, as Hurricane Harvey bears down: 'Good luck to everybody.'" CNBC. https://www.cnbc.
com/2017/08/25/hurricane-harvey-trump-tells-texans-good-luck.html.

458 Cummings, W. (2017, October 13). "Trump says he met with the 'president of the Virgin Islands.'" *USA Today*. https://www.usatoday.com/story/news/politics/onpolitics/2017/10/13/trump-says-he-met-president-virgin-islands/763910001.

459 Darden, J. T. (2017, October 13). "Federal disaster aid for Puerto Rico isn't foreign aid—but Trump acts that way." *Washington Post*. https://www.washingtonpost.com/news/monkey-cage/wp/2017/10/13/federal-disaster-aid-for-puerto-rico-isnt-foreign-aid-but-trump-acts-that-way.

460 Lin, S. (2020, August 21). "Trump threatens to withhold California wildfire funds. 'You gotta clean your forests.'" *The Sacramento Bee*. https://www.sacbee.com/news/nation-world/national/article245137275.html.

461 Frank, T. (2020, October 19). "Trump bypassed rules on Calif. disaster approval." *E&E News*. https://www.eenews.net/stories/1063716473.

462 Montano, S. (2020, January 22). "'Mops and buckets' won't do anything to save us from climate disaster." *The Week*. https://theweek.com/articles/890608/mops-buckets-wont-anything-save-from-climate-disaster?fbclid=IwAR0jcfXFrDmiKJYZ6FDjtN4-942ZwQVJ3YfHMDN5a6V1ZG_7sD3WddLHI8s.

463 Montano, S. (2017, August 28). "Here's when Trump's response to Hurricane Harvey will really matter." *Vox*. https://www.vox.com/2017/8/26/16208292/hurricane-harvey-trump-response.

464 Wan, W. and Miroff, N. (2018, September 15). "Alarm grows inside FEMA as administrator Brock Long fights for his job." *Washington Post*. https://www.washingtonpost.com/world/national-security/alarm-grows-inside-fema-as-administrator-brock-long-fights-for-his-job/2018/09/15/995fb280-b854-11e8-94eb-3bd52dfe917b_story.html.

465 Lind, D. (2018, September 12). "Is Trump using hurricane relief money to fund ICE? Not exactly." *Vox*. https://www.vox.com/policy-and-politics/2018/9/12/17850594/hurricane-fema-ice-trump-immigration.

466 Clark, D. (2018, September 12). "FEMA head defends $10 million transfer to ICE, accuses Democrat of 'playing politics.'" NBC News. https://www.nbcnews.com/politics/white-house/fema-head-defends-10-million-transfer-ice-accuses-democrat-playing-n908896.

467 Smith, A. (2019, September 1). "Trump officials defend decision to send FEMA funds to ICE ahead of Hurricane Dorian." NBC News. https://www.nbcnews.com/politics/donald-trump/trump-officials-defend-decision-send-fema-funds-ice-ahead-hurricane-n1048731.

How Not to Manage a Pandemic

468 Mosk, M. (2020, April 5). "George W. Bush in 2005: 'If we wait for a pandemic to appear, it will be too late to prepare.'" ABC News. https://abcnews.go.com/Politics/george-bush-2005-wait-pandemic-late-prepare/story?id=69979013.

469 DHS. (2018). "Threat and Hazard Identification and Risk Assessment (THIRA) and Stakeholder Preparedness Review (SPR) Guide." https://www.fema.gov/media-library/assets/documents/181470.

470 N.A. (2018). National Biodefense Strategy. https://www.hsdl.org/?abstract& did=815921.

471 Coats, D. (2019). "Worldwide Threat Assessment of The US Intelligence Community." https://www.dni.gov/files/ODNI/documents/2019-ATA-SFR---SSCI.pdf.

472 Blake, A. (2020, March 19). "Trump keeps saying 'nobody' could have foreseen coronavirus. We keep finding out about new warning signs." *Washington Post.* https://www.washingtonpost.com/politics/2020/03/19/trump-keeps-saying-nobody-could-have-foreseen-coronavirus-we-keep-finding-out-about-new-warning-signs.

473 Blake, A. (2020, March 19). "Trump keeps saying 'nobody' could have foreseen coronavirus. We keep finding out about new warning signs." *Washington Post.* https://www.washingtonpost.com/politics/2020/03/19/trump-keeps-saying-nobody-could-have-foreseen-coronavirus-we-keep-finding-out-about-new-warning-signs.

474 Nojang, E. N., and Jensen, J. (2020). "Conceptualizing Individual and Household Disaster Preparedness: The Perspective from Cameroon." *International Journal of Disaster Risk Science*, 1-14.

475 Toosi, N., Lippman, D., and Diamond, D. (2020, March 16). "Before Trump's inauguration, a warning: 'The worst influenza pandemic since 1918.'" *Politico.* https://www.politico.com/news/2020/03/16/trump-inauguration-warning-scenario-pandemic-132797.

476 Torbati, Y. and Arnsdorf, I. (2020, April 3). "How Tea Party Budget Battles Left the National Emergency Medical Stockpile Unprepared for Coronavirus." *ProPublica.* https://www.propublica.org/article/us-emergency-medical-stockpile-funding-unprepared-coronavirus.

477 N.A. (2020, March 25). "Partly false claim: Trump fired entire pandemic response team in 2018." Reuters. https://www.reuters.com/article/uk-factcheck-trump-fired-pandemic-team/partly-false-claim-trump-fired-pandemic-response-team-in-2018-idUSKBN21C32M.

478 Cameron, B. (2020, March 13). "I ran the White House pandemic office. Trump closed it." *Washington Post.* https://www.washingtonpost.com/outlook/nsc-pandemic-office-trump-closed/2020/03/13/a70de09c-6491-11ea-acca-80c22bbee96f_story.html.

479 Kessler, G., Rizzo, S., and Kelly, M. (2020, July 13). "President Trump has made more than 20,000 false or misleading claims." *Washington Post.* https://www.washingtonpost.com/politics/2020/07/13/president-trump-has-made-more-than-20000-false-or-misleading-claims.

480 Hetherington, M. and Ladd, J. M. (2020, May 1). "Destroying trust in the media, science, and government has left America vulnerable to disaster." Brookings. https://www.brookings.edu/blog/fixgov/2020/05/01/destroying-trust-in-the-media-science-and-government-has-left-america-vulnerable-to-disaster.

481 Lipton, E. (2020, April 11). "The 'Red Dawn' Emails: 8 Key Exchanges on the Faltering Response to the Coronavirus." *New York Times.* https://www.

nytimes.com/2020/04/11/us/politics/coronavirus-red-dawn-emails-trump.html.

482 For more on Kushner's involvement at FEMA see: Confessore, N., Jacobs, A., Kantor, J., Kanno-Youngs, Z., and Ferré-Sadurní. (2020, May 5). "How Kushner's Volunteer Force Led a Fumbling Hunt for Medical Supplies." *New York Times.* https://www.nytimes.com/2020/05/05/us/jared-kushner-fema-coronavirus.html.

483 Confessore, N., Jacobs, A., Kantor, J., Kanno-Youngs, Z., and Ferré-Sadurní. (2020, May 5). "How Kushner's Volunteer Force Led a Fumbling Hunt for Medical Supplies." *New York Times.* https://www.nytimes.com/2020/05/05/us/jared-kushner-fema-coronavirus.html.

484 Early research has found democratic governors were more likely to follow public health guidance on initial stay-at-home orders: Baccini, L. & Brodeur, A. (2020, December 1). "Explaining Governors' Response to the COVID-19 Pandemic in the United States." American Politics Research. https://journals.sagepub.com/doi/full/10.1177/1532673X20973453.

485 Knowles, S. G. (2020, April 21). "Kathleen Tierney—What Do Disasters Reveal About Society?" [Video]. COVID Calls. https://www.youtube.com/watch?v=54MjhyeqFWM&ab_channel=ScottGabrielKnowles-.

486 Svitek, P. (2020, July 2). "Gov. Greg Abbott orders Texans in most counties to wear masks in public." *The Texas Tribune.* https://www.texastribune.org/2020/07/02/texas-mask-order-greg-abbott-coronavirus.

487 Yong, E. (2020, November 20). "Hospitals Know What's Coming." *The Atlantic.* https://www.theatlantic.com/health/archive/2020/11/americas-best-prepared-hospital-nearly-overwhelmed/617156.

488 McEvoy, J. (2020, December 1). "1 In Every 800 North Dakota Residents Now Dead From COVID." *Forbes.* https://www.forbes.com/sites/jemimamcevoy/2020/12/01/1-in-every-800-north-dakota-residents-now-dead-from-covid/?utm_campaign=forbes&utm_source=facebook&utm_medium=social&utm_term=Gordie&fbclid=IwAR1693JyJ-XgI2DgAzBwZBd4btY3xblwdK7LOL3Fcc4a7WECKCqAQBpzdUU&sh=36ce83654555.

489 Shell, E. R. (2020, October 16). "How Straight Talk Helped One State Control COVID." *Scientific American.* https://www.scientificamerican.com/article/how-straight-talk-helped-one-state-control-covid.

490 Redlener, I., Sachs, J. D., Hansen, S., and Hupert, N. (2020). "130,000-210,000 Avoidable Covid-19 Deaths—And Counting—In the US." New York: National Center for Disaster Preparedness, Columbia University.

491 AP. (2020, September 10). "Public vs. private: A timeline of Trump's comments on virus." AP News. https://apnews.com/article/bob-woodward-virus-outbreak-donald-trump-archive-united-states-6d3932070522339fa91b8b3d3fc046fa.

492 Franck, T. (2020, February 28). "Trump says the coronavirus is the Democrats' 'new hoax'." CNBC. https://www.cnbc.com/2020/02/28/trump-says-the-coronavirus-is-the-democrats-new-hoax.html.

493 CDC. (2019, March 20). "1918 Pandemic." Center for Disease Control and Prevention. https://www.cdc.gov/flu/pandemic-resources/1918-pandemic-h1n1.html.

494 Chowkwanyun, M., and Reed Jr., A. L. (2020, July 16). "Racial health disparities and Covid-19—caution and context." *New England Journal of Medicine*. https://www.nejm.org/doi/full/10.1056/NEJMp2012910.

495 Wood, D. (2020, September 23). "As Pandemic Deaths Add Up, Racial Disparities Persist—And In Some Cases Worsen." NPR. https://www.npr.org/sections/health-shots/2020/09/23/914427907/as-pandemic-deaths-add-up-racial-disparities-persist-and-in-some-cases-worsen.

496 AP. (2020, June 21). "Maine has widest coronavirus racial disparity gap in the country." *Boston Globe*. https://www.bostonglobe.com/2020/06/21/nation/maine-has-widest-coronavirus-racial-disparity-gap-country/.

497 Koran, M. (2020, March 30). "Las Vegas parking lot turned into 'homeless shelter' with social distancing markers." *The Guardian*. https://www.theguardian.com/us-news/2020/mar/30/las-vegas-parking-lot-homeless-shelter.

498 Nelson, L. and Lau, M. (2020, December 18). "The wealthy scramble for COVID-19 vaccines: 'If I donate $25,000…would that help me?'" *Los Angeles Times*. https://www.latimes.com/california/story/2020-12-18/wealthy-patients-scramble-covid-19-vaccine.

499 *Pandemic Disaster Assistance Act of 2020*, S.3534, 116th Cong. https://www.congress.gov/bill/116th-congress/senate-bill/3534?q=%7B%22search%22%3A%5B%22Harris+Pandemic+Stafford%22%5D%7D&s=4&r=2.

500 Ridge, T. (2020, March 25). "In support of President's Trump use of the Stafford Act." *New York Daily News*. https://www.nydailynews.com/opinion/ny-oped-trump-stafford-act-20200325-lhpv7a6cqfdyrgf4qtaqzxp-6pu-story.html.

501 Song, L. and Yeganeh, T. (2020, April 29). "Grieving Families Need Help Paying for COVID-19 Burials, but Trump Hasn't Released the Money." *ProPublica*. https://www.propublica.org/article/fema-burial-relief-coronavirus.

502 This history of the flood is from: Barry, J. M. (2007). *Rising Tide: The Great Mississippi Flood of 1927 and How It Changed America*. Simon & Schuster.

PART SIX

Disaster Justice

503 This account of Love Canal comes from Gibbs, L. M. (2011). *Love Canal: and the Birth of the Environmental Health Movement*. Island Press.

504 For more on Love Canal activism including a race, class, and gender analysis read: Blum, E. D. (2008). *Love Canal Revisited: Race, Class, and Gender in Environmental Activism*. Lawrence: University Press of Kansas.

505 Hautzinger, D. (2020, February 24). "The Chicago Woman Who Fought

to Clean Up the Southeast Side." WWTW. https://interactive.wttw.com/playlist/2020/02/24/hazel-johnson.

Creating a Movement for Disaster Justice

506 Duara, N. (2016, September 8). "National Guard Activated as Protesters Await Pipeline Decision." *Los Angeles Times.* https://www.latimes.com/nation/la-na-pipeline-national-guard-20160908-snap-story.html.

507 Carasik, L. (2016, November 21). "N. Dakota Pipeline Protest Is a Harbinger of Many More." *Aljazeera.* https://www.aljazeera.com/opinions/2016/11/21/n-dakota-pipeline-protest-is-a-harbinger-of-many-more.

508 Thorbecke, C. (2016, November 3). "Why a Previously Proposed Route for the Dakota Access Pipeline Was Rejected." ABC News. https://abcnews.go.com/US/previously-proposed-route-dakota-access-pipeline-rejected/story?id=43274356.

509 For more on the legal history of the pipeline see: White, G. W., and Millett, B. V. (2019). "Oil Transport and Protecting Clean Water: The Case of The Dakota Access Pipeline (DAPL)." *Present Environment and Sustainable Development* (2), 115-128.

510 For a timeline of the Standing Rock protests see: Javier, Carla. (2016, December 14). "A Timeline of the Year of Resistance at Standing Rock." *Splinter News.* https://splinternews.com/a-timeline-of-the-year-of-resistance-at-standing-rock-1794269727.

511 Elbein, S. (2017, January 31). "The Youth Group That Launched a Movement at Standing Rock." *New York Times Magazine.* https://www.nytimes.com/2017/01/31/magazine/the-youth-group-that-launched-a-movement-at-standing-rock.html.

512 For a contextualized history of the protests in the history of Indigenous resistance see: Estes, N. (2019). *Our History Is The Future: Standing Rock Versus The Dakota Access Pipeline, and The Long Tradition of Indigenous Resistance.* Verso Books.

513 For more on the Standing Rock Protests see: Estes, N., and Dhillon, J. (Eds.). (2019). *Standing With Standing Rock: Voices From The #NoDAPL Movement.* University of Minnesota Press.

514 For a history on Lakota media coverage: Monet, J. (2017). "Covering Standing Rock." *Columbia Journalism Review.* https://www.cjr.org/local_news/covering-standing-rock.php.

515 For an analysis of different types of media coverage see: Hunt, K., and Gruszczynski, M. (2019). "The influence of new and traditional media coverage on public attention to social movements: the case of the Dakota Access Pipeline protests." *Information, Communication & Society,* 1-17.

516 Barajas, J. (2016, November 21). "Police deploy water hoses, tear gas against Standing Rock protesters." PBS News Hour. https://www.pbs.org/newshour/nation/police-deploy-water-hoses-tear-gas-against-standing-rock-protesters.

517 Wong, J.C. (2016, November 22). "Dakota Access pipeline protester 'may

lose her arm' after police standoff". The Guardian. https://www.the-guardian.com/us-news/2016/nov/22/dakota-access-pipeline-protester-seriously-hurt-during-police-standoff-standing-rock.

518   Sammon, A. (2016, December 4). "How Did Police From All Over the Country End Up at Standing Rock?" *Mother Jones.* https://www.motherjones.com/environment/2016/12/standing-rock-police-militarized-emergency-management-assistance-compact-north-dakota.

519   Rott, N. (2016, December 4). "In Victory For Protesters, Army Halts Construction of Dakota Pipeline." NPR. https://www.npr.org/sections/thetwo-way/2016/12/04/504354503/army-corps-denies-easement-for-dakota-access-pipeline-says-tribal-organization.

Working Toward Disaster Justice

520   AP. (2006, April 24). "Nagin fails to avoid runoff." *Deseret News.* https://www.deseret.com/2006/4/24/19949828/nagin-fails-to-avoid-runoff.

521   For more about this idea in the broader context of the climate crisis see: Heglar, M. A. (2020, April 4). "We Can't Tackle Climate Change Without You" Wired. https://www.wired.com/story/what-you-can-do-solve-climate-change/.

On Finding Courage

522   Marvel, K. (2018, March 1). "We Need Courage, Not Hope, to Face Climate Change." *On Being.* https://onbeing.org/blog/kate-marvel-we-need-courage-not-hope-to-face-climate-change.

Epilogue

523   Henson, B. (2020, October 17). "Iowa derecho in August was most costly thunderstorm disaster in U.S. history." *Washington Post.* https://www.washingtonpost.com/weather/2020/10/17/iowa-derecho-damage-cost.

Index